Ma

System Architecture of Small Autonomous UAVs

Marek Musial

System Architecture of Small Autonomous UAVs

Requirements and Design Approaches in Control, Communication, and Data Processing

VDM Verlag Dr. Müller

Imprint

Bibliographic information by the German National Library: The German National Library lists this publication at the German National Bibliography; detailed bibliographic information is available on the Internet at http://dnb.d-nb.de.

Cover image: www.purestockx.com

Publisher:
VDM Verlag Dr. Müller Aktiengesellschaft & Co. KG , Dudweiler Landstr. 125 a, 66123 Saarbrücken, Germany,
Phone +49 681 9100-698, Fax +49 681 9100-988,
Email: info@vdm-verlag.de

Zugl.: Berlin, Techn. Univ., Habil., 2007.

Produced in USA and UK by:
Lightning Source Inc., La Vergne, Tennessee, USA
Lightning Source UK Ltd., Milton Keynes, UK
BookSurge LLC, 5341 Dorchester Road, Suite 16, North Charleston, SC 29418, USA

ISBN: 978-3-8364-6113-9

Abstract

With the miniaturization of processing units, sensors, and communication devices rapidly progressing, interest in small autonomous *unmanned aerial vehicles* (UAVs) is currently expanding exponentially. This thesis provides a survey of the requirements and suitable solution approaches that are typically connected with the system architecture of small autonomous UAVs. Remarkably, there is not as much variation in the set of these requirements as one might at first glance expect. Any autonomous UAV needs to provide means of communication with its human users on the ground, to be able to measure its position and attitude, to move to some position as required by the users, and often to perform certain measurements and/or manipulation there. Naturally, all of these tasks involve some on-board means of information processing. Finally, the smaller the UAV, the stricter will be the limitations imposed on the weight and power consumption of the components employed, and the more costly will inappropriate design decisions turn out.

This thesis describes these typical requirements thoroughly and presents solutions, sometimes a number of alternative ones, that have been successfully applied in actually working small autonomous UAV systems. As the author has been deeply involved in the design of TU Berlin's helicopter UAV *MARVIN*, this system serves to provide particularly detailed insights into possible and proven solutions.

Acknowledgements

This thesis constitutes the resume of more than 13 years (oh, really?) of work in the field of unmanned aerial vehicles. During this time, one thing has proven itself repeatedly: Being successful in this field requires the cooperation of many individuals, due to the enormous range of issues to be addressed, work to be done, and things to be watched "in the field". Therefore, I wish to express my deepest gratitude to all members of the TubRob and MARVIN teams, who have contributed solutions, motivation, and lots of good spirit that have all been crucial to the accomplishment of this thesis. While it is quite impossible to include all of their names, I want to mention at least Marion Finke and Wolfgang Brandenburg, who were initially responsible for my getting hooked up with UAVs, and pick the remaining members of our IARC final team of 2000, who where Marc Bartholomäus, Eike Berg, Carsten Deeg, Christian Fleischer, Christian Reinicke, Volker Remuß, Andreas Rose, Roland Stahn, and Andreas Wege. Very special thanks to Matthias Jeserich, our safety pilot, who ingeniously saved more material than we could have ever fixed again, for example on the very day of the competition final. More recently, Volker Remuß and Carsten Deeg have been most deeply involved in the further development of the MARVIN system, and I am grateful for the pleasure and productivity of the work we have been sharing, and for their indispensable contributions in their respective fields of expertise. Thanks to Yorck Rackow for bravely – and successfully – following Matthias as the safety pilot of MARVIN Mark II. Some special thanks go to Spain, for a very pleasant and resourceful cooperation and lots of happy dining, to Antonio, Jorge, Juan Carlos, Roberto, and Theo.

I highly appreciate the work of Aníbal Ollero from Sevilla, who has done a wonderful job in organizing the successful COMETS project and, hence, has contributed a lot to both advancing and promoting our group's work. So it was almost inevitable to ask him to join the circle of examiners for this thesis, which he generously accepted.

The one particular person who has always favored my UAV work, and has always provided me with just the right mixture of material support, deep confidence, and the freedom of choosing my modus operandi myself, is Günter Hommel, head of the Real-Time Systems group at TU Berlin. My most sincere thanks for these wonderful conditions, and for acknowledging both theoretical and practical results in a research area that does inevitably involve incredibly dangerous rotor-blades and quite heavy things that might actually fall straight onto your head, from 50 m out of the sky. Many thanks to him as well as to Prof. Ollero for examining so much stuff that is in fact much more interesting to people who actually want to build a UAV in person.

More thanks to the remaining members of the habilitation committee, Prof. Jörg Raisch and Prof. Olaf Hellwich, for their time and for taking care of the most official parts. Finally, Gudrun Pourshirazi deserves very special thanks for her patience in guiding us through paperwork and always being somebody to just happily rely upon.

Contents

List of Figures

List of Tables

17

Chapter 1

Introduction

Within the huge and exceptionally fast evolving research area of autonomous mobile robots, the class of autonomously *flying* robots, or *unmanned aerial vehicles* (UAVs), has ever since its appearance played a somewhat special role. There are numerous reasons for its specialty quite easily found, as compared to robot vehicles operating on the ground or in the water:

- Airborne locomotion provides some difficulty in itself, completely irrespective of the vehicle's autonomy. It consumes relatively much energy and imposes stringent weight limitations. Furthermore, navigation in UAVs involves up to six instead of usually just three degrees of freedom (e.g. surface position plus yaw orientation in the case of surface-based vehicles).

- There is usually no easy fail-safe mode for UAVs. Just switching off all actuators will typically *not* result in a safe state of the UAV once airborne, due to the substantial amount of energy stored in its movement and altitude over ground. This requires the design of dedicated fail-safe procedures and special attention to all aspects of reliability, and it tends to assign particular bodily and monetary risk to any operation of UAVs.

- Due to the kinematics of the UAV and/or its propulsion, its operation is often restricted to outdoors. This makes any kind of testing with hardware involved much more tedious and time-consuming than with surface-based robots. Therefore, system development in the case of a UAV usually depends on the existence of a simulation of its dynamics.

- The UAV's flight control usually requires a 3D position sensor and, in most cases, also a 3D orientation sensor. Odometry, as used in many ground-based robots, is not an option, due to the loose coupling with the environment and much higher accuracy requirements. In many cases, the UAV's being equipped with a GPS receiver primarily for position measurement, and acceleration, rotation, and magnetic field sensors primarily for orientation measurement is the *minimum* possible solution.[1] At the same time, the use of GPS will also restrict the UAV's operation to outdoors.

---

[1]For fixed-wing class UAVs, suitably relaxed requirements to their navigation accuracy may render orientation measurement redundant, or at least permit to drop some of the sensors mentioned.

These exceptional characteristics are responsible for the fact that the field of autonomous UAVs has yet remained relatively small, however, with regard to the number of autonomous robot systems designed altogether. Nevertheless, with the current state-of-the-art in miniaturizing processing units, sensors, communication devices, and energy storage, the frontier of possible small autonomous UAVs has entered a rapid progress, as well as interest in them and applications held possible. The first truly autonomous UAVs, for military applications, have been put into use as early as in the 1960s. A renowned pioneer system with exceptional success is constituted by the Ryan Model 147 series UAVs, with detailed descriptions of its history and variants available on the web in [59, 141]. Much later, DARPA's Micro Air Vehicle (MAV) Program initiative (see e.g. [61]) launched in 1996 more or less marks the "public" birth hour of the idea of using very small UAVs.

This book covers only "small" UAVs, setting an admittedly arbitrary limit at 20 kg take-off mass. Bigger vehicles tend to be reserved to the military sector, to only partly fulfill the special characteristics as enumerated above, and to smoothly pass into the field of "conventional" autopiloted planes in every respect. Besides, 20 kg is the most relevant legal size limit for model aircraft in Germany, and a quite reasonable upper limit for UAVs to be developed with model aircraft technology and procedures anywhere around the world.

The term "autonomy" in the sense of this book shall refer to the UAV's ability to navigate towards at least one desired position, or to follow some desired trajectory in the form of at least a straight line, independent of any further steering commands from the ground station or any other off-board sources. Basically, this could be easily formalized by requiring the UAV to be fully operational without any continuous information exchange to or from the UAV. However, for the maximum accuracy of certain GPS position sensors, a stream of reference data must be continuously supplied to the UAV (*differential GPS*, DGPS). Therefore, the exact convention throughout this book will be to define autonomy as the *absence of any closed-loop information exchange between the UAV and external parts of the UAV system*. This convention will perfectly permit the use of DGPS, while it reliably prohibits the off-board implementation of any part of flight control.

The subject-matter of this book is constituted by multiple aspects of the system architecture of such small autonomous UAV systems. Regarding each of these aspects, requirements will be collected and successfully applied solutions will be set forth. This structure is quite expedient, for it turns out that both the requirements and possible approaches to system architecture do not vary too much over a wide range of different kinds of UAVs. Instead, it is quite possible to identify *typical* requirements and *typical* solutions or at least classes thereof.

Many reasons for this limited variance have already been mentioned in the list of characteristics above: strict weight and power limitations, limited choice in position and attitude sensors, and inherent risks. Additionally, UAV systems are particularly distinct occurrences of distributed systems, inducing challenging demands with respect to the communications infrastructure between, at least, one UAV and the corresponding ground station. These demands relate to the magnitude and multitude of data to be exchanged as well as to the strict real-time constraints to be met. Furthermore, due to the dynamics of aerial vehicles, there is even a certain amount of common structure of the flight controller, which always has to perform 3D position control in some way, which in turn requires some sort of subordinate attitude control for any usual kind of aircraft (e.g. airship, helicopter, fixed-wing plane, or some multi-rotor vehicle). Finally, sensor data processing, ground station interaction, and flight control require, at least, one on-board computer running different software modules according to some real-time-enabled processing scheme.

All in all, it is supremely useful to collect and systematize these common aspects. This book tries to accomplish this task, in order to provide a good overview of relevant problems and solutions in the field of UAV system design, and to spare "novice" UAV designers some errors already made by others.

The remainder of this introduction will postpone the list of aspects of small UAV system architecture to be covered by this book to the overview section finishing the introduction. The immediately following section deals with possible application scenarios of small autonomous UAVs at first. Next, there is an overview of research centers most visibly engaged in small UAV development, followed by a very basic outline of air vehicle flight controls.

1.1 Applications

In principal, there is an almost unlimited number of applications imaginable for small autonomous UAVs. This section provides an overview – naturally, an incomplete one – of application scenarios. These scenarios will *not* be analyzed in detail, because economical issues and technical issues of different fields of application are outside the scope of this book.

In the following, application scenarios are classified hierarchically, first by the basic type of task (section 1.1.1), the type of sensors employed (section 1.1.2), and the purpose of the application in question (section 1.1.3).

1.1.1 Task Type

The classification by *task types* shall basically refer to the following distinction:

Sensing In tasks of the *sensing* type, there is no effect on the environment intended. The only purpose is the acquisition of data, which may be recorded on-board the UAV or transmitted to a ground station.

Manipulation In the *manipulation* case, some effect on the environment constitutes the primary objective. Of course, manipulation tasks will often require concurrent sensing in order to control the process of manipulation.

Relaying The use of a UAV as a communications relay shall be regarded separately. This means that the UAV receives and retransmits data from some external source to some external destination. While relaying could in some way be regarded as a combination of sensing and manipulation, it is distinguished here because none of the endpoints involved is "environment", but they are parts of the system the UAV belongs to.

The great majority of possible application scenarios are focused on sensing, while manipulation tasks are of limited eligibility in general. The reasons for this are

- the small payload, which directly results from the 20 kg take-off mass limit introduced above,

- the higher complexity of manipulation tasks, which makes their automation much less feasible, or at least confines the application of UAVs to highly specialized environments and/or tasks.

1.1.2   Sensor Type

This section lists the types of sensors that might be desirable for small UAV sensing applications.

Optical   Certainly, the most attractive sensor type for use on-board a UAV are cameras for visible light images or video streams. In addition to fulfilling some observation task, prompt image transmission can also serve as a valuable support for the human operator on the ground, to assist the UAV's operation and to decide about its mission's progress. There is a big variety of cameras available that can be easily carried by a small UAV.

Infrared   Infrared cameras are, of course, only a special case of optical sensors. It has to be noted that most high-performance IR cameras are unsuitable for small UAV use because of their integrated cooling devices. Infrared images are superior to visible-light ones regarding the detection of fire and of humans or animals, for example.

Chemical   With chemical sensors, a UAV can detect the presence of certain substances or determine their concentration in the surrounding air. Together with a suitable navigation strategy, it might even be possible to locate the source of the substance in question.

Weather   Weather sensors shall subsume all sorts of physical sensors measuring some quantity related to the atmosphere. That is, temperature, air pressure, humidity, wind, visibility, etc. Weather observation is an attractive UAV application because of its possible long observation time and its rare need of human intervention. The well-known registering balloon constitutes, after all, some sort of UAV as well.

Magnetic/Electric Field   Sensors for magnetic and electric fields may be used in certain geological or archaeological applications as well as for the detection of specific objects (e.g. mines).

1.1.3   Application Classes

In this section, potential application cases are listed. They are grouped into a number of classes, each of which is presented within a single subsection. Again, most of the scenarios mentioned are sensing applications (see 1.1.1).

Facility Monitoring

Facility monitoring shall denote the occasional or periodical surveillance of the operational reliability of some technical facility. UAV monitoring will be economically attractive especially
in cases of distributed facilities. These kinds of facilities include:

Traffic Infrastructures, such as railroads or possibly cable-cars.

Power Transmission Lines, which today are routinely inspected through manned helicopter
flights to determine the state and detect possibly necessary repairs of cables and insula-
tors.

Pipelines for oil and gas transportation need to be periodically checked for leakage. Today,
this is usually done via manned helicopter flights as well.

Cattle Observation is an application that has already been researched. The task in this case
is to monitor the flock size and the exact whereabouts of the animals with respect to
certain zones of the farm area.[2]

In all these cases, the profitability of using small UAVs depends on the necessary degree of
human intervention during the monitoring process, on the required operation range of the UAV
without landing, and on the UAV's traveling speed as compared to that of a manned helicopter.

Human Surveillance

Human surveillance shall subsume all observation tasks that relate to the presence, activity,
or mobility of people, with or without vehicles. This results at least in the following specific
application scenarios:

Monitoring of Events, such as concerts or all sorts of celebrations, especially for security
reasons. Aerial imaging may provide affordable high-quality image data from variable
angles. Additionally, it may exert some desirable deterrent effect – as long as small
UAVs are not yet small enough to pass unnoticed, at least. A particularly big problem
in this area is, of course, the aspect of the UAVs' operational safety, with respect to the
people being monitored.

Border Patrol is a "classical" human surveillance task especially suited for UAV employment, due to the size of the region to be observed and the simplicity and monotonicity
of this task. Of course, this mainly applies to more or less "closed" borders, where the
pure presence of somebody is already a criterion to raise an alarm.

Plant Security may be seen as a scaled-down case of border patrol. The larger the plant to
be monitored is, the more attractive will the use of UAVs be.

---

[2]The classification of this task under *facility* monitoring may seem debatable, but has been chosen due to the
industrial nature of cattle breeding.

Traffic Monitoring also belongs, according to the definition of human surveillance from above, into this class of tasks. It is one of the easier tasks of the human surveillance group and can be performed from a number of predefined positions that may easily be in safe distance from the observed traffic facilities. It may be fully automated, since it mainly consists in counting with very little amount of "decision".

Environmental Monitoring

Environmental monitoring shall refer to all measurement and observation tasks regarding inartificial systems and phenomena. Potential scenarios of this class include the following:

Fire Detection and Observation may constitute a highly relevant task, for example in southern Europe and in the United States, but even in parts of Germany. Today, the task of fire detection is often performed by observation towers – sometimes manned, sometimes carrying cameras. The use of small UAVs would provide wider view angles and more flexibility as soon as detection transforms into observation. In the EC Fifth Framework Programme's project COMETS [28], the observation of forest fires was chosen as a sample application to be performed by a fleet of small UAVs.

Geological Monitoring shall subsume the observation of events and items like landslides, debris flows, glaciers, avalanches, geysers, or volcanic activity. While in several of these cases there may not be great benefits earned from the employment of UAVs as compared to surface-based measurement equipment, air-based observation might prove beneficial in others. A very special case of geological monitoring might be found in artificial full-scale debris-flow experiments, as have been carried out for more than 30 years in the Zailisky Alatau Mountains near Almaty, Kazakhstan. Here, the timing of the events of interest is known beforehand, and the nature of debris-flows involves movement, which renders the use of mobile cameras particularly attractive. The background is explained e.g. in [70], yet the paper focusses on sensors deployed *within* a debris flow.

Climate Observation is a very attractive UAV application, due to its simplicity and the low weight of the required sensors. The recording of atmospheric data along a predefined flight trajectory is even easier performed using an auto-piloted vehicle than with a "man-in-the-loop" setting. There have already been successful applications of remotely piloted model aircraft, e.g. for wind system research in the Andes [119].

Animal Observation, for research purposes, may profit from the use of UAVs due to the combination of mobile cameras with a comparably low noise signature. A prime application would be the imaging of flocking birds, as done for example for the 2001 movie *Winged Migration* by Jacques Perrin, where the filming was done via manned ultralight planes, hang gliders, and balloons, however[3].

---

[3]This exact application might as well belong into the class of entertainment applications, though, i.e. 1.1.3.

Entertainment

Entertainment applications primarily pertain to movie and television shooting. In particular, UAVs could be effectively used for the following:

Sports Programs could be suitably recorded by UAVs, because sports competitions tend to follow a predefined track, and it is even imaginable to have competitors carry e.g. GPS receivers in order to obtain the current position of every competitor. Of course, special care needs to be taken not to endanger both competitors and spectators by the use of UAVs, which should also be possible due to the known structure of the track and spectators' areas. All sorts of outdoor competitions may be eligible here, like marathon runs, cycling, skiing, sailing, wind surfing, or car racing, just as a few examples.

Movie Shooting, in the case of expensive special-effect scenes with high involvement of technical components, may profit from preprogrammed aerial camera trajectories. This is due to the determinism and repeatability of the UAV's movement. Another application in movie production would be the autonomous operation of model aircraft for "fake" air combat scenes and the like. The use of remotely piloted scale aircraft for this purpose is quite common today. Of course, the frequency of both kinds of scene shooting is currently reducing in favor of computer graphics.

Event Programs are just a generalization of sports programs, as discussed above. Concerts, exhibitions, and large-scale artwork are possible examples of non-sports events that might be subject to UAV-based imaging, too.

Documentation

The class of *documentation* tasks shall refer to static, i.e. "non-event" documentation applications. Possible examples are:

Architecture documentation, through close-up imaging of buildings, can hardly be performed in any other way. Consider the task of generating a virtual reality model of the exterior of some famous church, for example. In this case, it is desirable to obtain close-up images of the whole of the building's surface area, at least in order to generate accurate textures for some real-looking rendering of the outside. This may easily be impossible from the ground due to various kinds of obstruction, and it is certainly impossible via manned aircraft due to legal and safety reasons. Therefore, this constitutes an application area to be initiated through the use of small UAVs.

Geologic modeling and mapping tasks, similar to the ones in architecture, may also benefit from UAV imaging. Especially, 3D mountain terrain mapping and texturing may as well be something mostly impossible (or much more expensive) by any other means.

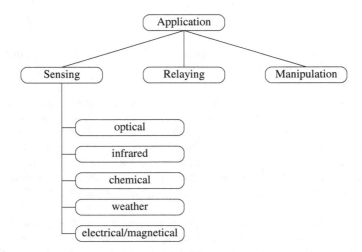

Figure 1.1: Categorization of small UAV applications by task and sensor types.

Disaster Response

Due to the inherent dangers and relative urgency associated with different kinds of disaster scenes, the use of UAVs by *disaster response* forces is particularly attractive and has already received considerable attention of the research community. Specific kinds of disaster response applications may focus, amongst others, on the following kinds of disasters:

Earthquakes of high destructive power usually concern a larger area, so that the use of UAVs for assessing the damage and locating survivors may be attractive.

Chemical Accidents pose a potentially high risk for the response forces. Therefore, the combination of visual and chemical sensors in a small UAV constitutes a good tool to evaluate the situation after accidents in chemical plants or similar facilities.

Nuclear Accidents are another reasonable arena for the use of UAVs, for reasons pretty similar to the ones given above for chemical accidents. In the case of nuclear accidents, the problem of contamination and the size of the area threatened might be much bigger. In particular, it might be reasonable to employ UAVs as disposable items, in order to solve the problem of decontamination.

Floodings, like for example the great tsunami flooding in Asia in 2004, would as well be suitable for UAV disaster response applications. In this case, the surface-based reachability of parts of the concerned area may be substantially affected, so that the application of UAVs can generate a valuable speedup of the damage assessment procedure.

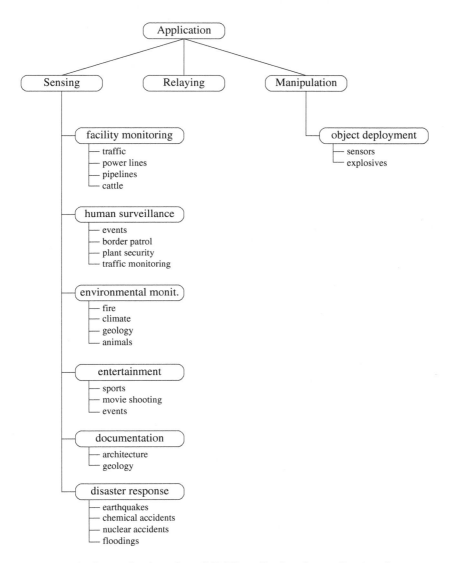

Figure 1.2: Categorization of small UAV applications by application classes.

Object Deployment

While all the application classes listed to this point have been sensing tasks according to the categorization of section 1.1.1, *object deployment* is one of the few reasonable application classes of the *manipulation* category. Specific examples include the following:

Deployment of Sensors constitutes one part of the use of sensor networks, which is in turn a highly fashionable research area currently. [41] is one of the primary launching papers of this topic.

 As those smart sensors are usually expected to be cheap and light-weight, it seems feasible to have them deployed via small UAVs. This way of deployment is particularly appealing from a research perspective. Furthermore, the UAVs could even serve as components of the sensor network after its deployment, for relaying purposes.

Deployment of Explosives, in particular for the purposeful triggering of avalanches, is a very specific application. However, being a structured but dangerous task in a known environment and considering the huge demand for people and equipment engaged in performing it, it seems to be a particularly well-suited kind of application.

1.1.4   Summary

Figure 1.1 summarizes the categorization of small UAV applications by task types and sensor types, as explained in sections 1.1.1 and 1.1.2. Figure 1.2 summarizes the categorization by application classes, as detailed in section 1.1.3.

1.2   Research Centers

This section provides a brief overview of major research centers notably engaged in small UAV research. Companies and other organizations with purely economical objectives, usually in the military segment, have been left out here deliberately.

1.2.1   University of California, Berkeley

The *Berkeley Aerobot Team* (BEAR) [118] at the University of California, Berkeley has been engaged in both fixed and rotary wing UAVs since 1996. It is headed by Shankar Sastry and belongs to the Department of Electrical Engineering and Computer Sciences (EECS). Current research topics include formation flight and vision-based navigation. However, some of the vehicles in use (Yamaha R50 and Yamaha RMAX helicopters) do exceed the definition of *small* UAVs as used within this book.

1.2.2 LAAS/CNRS, Toulouse

The *Laboratoire d'Analyse et d'Architecture des Systèmes* of the French *Centre National de la Recherche Scientifique* (LAAS/CNRS) [25] conducts the EDEN project concerned with *robotics in natural environments*. In this project, among other platforms, the airship robot *Karma* is used, mainly for visual and infrared environment monitoring tasks. The Robotics and AI group of LAAS, harboring the EDEN project, is lead by Raja Chatila. Karma is one of the vehicles that participated in the COMETS EC project [28].

1.2.3 DLR, Braunschweig

The *Systems Automation Department* at *German Aerospace Center* (DLR) works on autonomous VTOL (vertical take-off and landing) UAVs within the ARTIS project [39], conducted in Braunschweig. The Systems Automation Department is headed by Franck Thielecke. This ARTIS project is specifically targeted at providing a universal testbed and demonstrator platform to facilitate practical experiments with advanced UAV sensing and control approaches. The development of an autonomous UAV as such does not constitute a primary goal.

1.2.4 University of Texas, Arlington

The *Autonomous Vehicles Laboratory* (AVL) [116] at the University of Texas, Arlington is working on fixed-wing small UAVs, mainly for reconnaissance missions. It is headed by Arthur A. Reyes. In 2005, this lab's teams won the most recent *International Aerial Robotics Competition* (IARC) [46] held by the *Association for Unmanned Vehicle Systems International* (AUVSI) [45]. In this multi-year competition, UAVs are finally supposed to approach a target zone, identify a target building via light marks, and enter the target building in order to deliver visual reconnaissance data from inside.

1.2.5 Georgia Institute of Technology

The *Georgia Tech UAV Research Facility* (UAVRF) [78] at the Georgia Institute of Technology is headed by Eric N. Johnson. It was established in 1997. Research topics include flight control and UAV system software. Vehicles in use are of both fixed wing and rotary wing type. The UAVRF also participated in the IARC (see above) until 2003.

1.2.6 ETH Zürich

The *UAV Group* at the Measurement and Control Laboratory at ETH Zürich [55] has been researching small UAV systems since 1993. This includes fixed-wing planes, an airship, and – mainly – helicopters. A wide variety of topics have been investigated, including indoor-flight, acrobatic flight, assisted remote piloting, and universal controllers. ETH Zürich won second place in the 1996 IARC (see above). The essential research results can be found in the doctoral theses of Christoph Eck [40], Oliver Tanner [131], and Martin F. Weilenmann [139]. As a spin-off of the ETH Zürich research team, the company *weControl* was founded in 2000, providing the first commercial autopilot system for universal use in small (i.e. model scale) helicopters in Europe [138].

1.2.7 TU Braunschweig

The Institute of Aerospace Systems at TU Braunschweig, headed by Peter Vörsmann, conducts several projects on micro aerial vehicles [137] of fixed-wing type. The smallest vehicles studied are below 50 cm of wingspan. Specific research activities focus on minimization and fully autonomous navigation capabilities.

1.2.8 Technische Universität Berlin

The Real-Time Systems Group (PDV) at TU Berlin, headed by Günter Hommel, entered the development of small UAVs in 1993. After an initial airship-based vehicle [23, 22] for the 1995 IARC (2nd place), PDV focussed fully on helicopter UAVs [68]. The most relevant results were the winning of the IARC three-years-final in 2000 [101, 99, 103] and the successful participation in the EC COMETS project [28]. Currently, PDV investigates the coordinated control of multiple helicopter UAVs for joint transportation tasks, funded by the German Research Foundation (DFG), and the use of UAVs for the deployment and operation of sensor networks in the EC AWARE project [6].

1.3   Outline of Vehicle Controls

This section will provide a very rough and informal outline of the control surfaces and signals required to stabilize, control, and maneuver the most common types of air vehicles. Readers with only minimal familiarity with these vehicle mechanics are encouraged to safely skip this section, but others may benefit from this information later in the text. Vehicles addressed in the following subsections are airships, fixed-wing planes, and single-rotor helicopters. More detailed and fully formal treatments may be found elsewhere, e.g. in [38, 36, 80] for helicopters and [81] for planes.

1.3.1   Airships

Airships are characterized by the fact that the required lift is predominantly provided as static lift through some gas – usually helium – with a lower density than air. Airships, in contrast to balloons, additionally exhibit some motors and control surfaces for maneuvering. The equipment with control surfaces and, possibly, with a thrust vector control mechanism varies between actual vehicles, but the minimal configuration to be examined here is depicted in

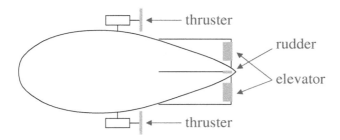

Figure 1.3: Flight controls outline of an airship-type vehicle.

Control	Type	Effect
thrust	propeller or jet	propulsion force
rudder	control surface	yaw torque
elevator	control surface	pitch torque

Table 1.1: Summary of typical airship controls.

figure 1.3: In this figure and the following ones for different kinds of vehicles, the vehicle is always seen from above, and control surfaces and devices are indicated through hatched areas. The airship is propelled by one or more engines producing a variable amount of *thrust* (here, it shall be assumed that all thrusters are always controlled together, so that the propulsion force generated is variable but no torque is produced). By placing its center of gravity beneath its center of lift, the airship's attitude is passively stabilized and does not require active control. At the tail, there is an empennage providing two sets of control surfaces, namely the *rudder* and the *elevator*. The rudder is used to turn the vehicle around the vertical axis (yaw), i.e. for flying curves. The elevator serves for changing the vehicle's pitch angle, which in turn allows to control its altitude.

Every airship requires a preflight *lift trim*, which is preferably slightly negative (downward). For it is somewhat inappropriate to end up with excessive lift in case of engine failure – although this may be alleviated by providing safety valves for the gas fill, of course. This negative static trim can then be compensated by a slight permanent "nose up" pitch angle during flight, which generates additional lift through the tilted thrust vector and the aerodynamics of the hull.

The control surfaces are of course only effective at a sufficient airspeed of the vehicle. Thus, while an airship does not technically depend on airspeed for remaining airborne, it does so effectively for maintaining its maneuverability. Due to the substantial hull size required to hold enough lifting gas, an airship's airspeed is usually limited and may in some cases become adversely dominated by the speed of wind. On the other hand, only sufficient wind permits hovering at zero ground speed without loss of control.

Table 1.1 summarizes the available airship controls explained above and their effects.

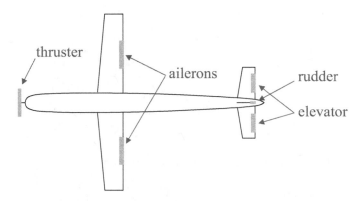

Figure 1.4: Flight controls outline of a fixed-wing-plane-type vehicle.

Control	Type	Effect
thrust	propeller or jet	propulsion force
rudder	control surface	yaw torque
elevator	control surface	pitch torque
aileron	control surface	roll torque

Table 1.2: Summary of typical fixed-wing-plane controls.

1.3.2   Fixed-Wing Planes

Fixed-wing planes are characterized by the fact that they produce lift through airfoils stati-
cally attached to the fuselage and sufficient airspeed. Figure 1.3 depicts the most traditional
configuration of a fixed-wing plane with respect to flight controls: The *thrusters*, usually one
or more propellers or jet engines, are primarily responsible of maintaining sufficient airspeed
and controlling it. Additionally, an increase in airspeed results in an increase in the lift generated by the wings, so that the thrusters also play a relevant role in altitude control. The control
surfaces provided within the plane's empennage basically correspond to the ones discussed in
the airship case – see above. But the rudder would not be sufficient for curve flying in the case
of a fixed-wing plane due to its typically higher velocity together with the smaller aerodynamic signature of its fuselage. Therefore, the centripetal force required for curve flying must
be generated by attaining a certain roll angle, which in turn produces a horizontal component
of the lift vector. This requires controls for the roll angle, which are provided in the form of
control surfaces at the wings that are always operated in opposing directions (up/down), thus
providing a roll torque. These surfaces are called *ailerons*.

The roll and pitch attitudes are usually inherently stabilized to some degree through particular
design details of a plane:

- The center of mass is usually located ahead of the wings' center of lift, with the horizontal tail producing negative lift to compensate. As a result, both the wings and the

horizontal tail counteract any pitching rotation, resulting in the passive stabilization of the latter.

- Unintended roll rotation leads to the kind of sideward airflow at the vertical tail that counteracts the roll rotation, due to the fact that the fin is above the plane's center of mass. This tends to passively stabilize roll rotation.

Curve flying is mainly initiated and controlled through the ailerons. During a curve, the rudder is only used to keep the fuselage's orientation aligned with the airflow, minimizing any undesired sideward attack of the airflow toward the fuselage (called *slip*). This effectively requires coordinated employment of ailerons and rudder. Climbing and declining is mainly accomplished via thrust control, for it is usually desirable to keep the airspeed as constant as possible, and climbing or declining primarily means increasing or reducing the plane's potential energy.

Table 1.2 summarizes the available plane controls explained above and their effects.

1.3.3 Helicopters

Single-rotor helicopters exhibit a single main rotor for lift generation, and a second, horizontal thruster – usually the tail rotor – that compensates for the yaw torque produced by the main rotor's rotation. Attitude and motion control of helicopters is somewhat more involved, for several reasons:

- A helicopter is inherently unstable with respect to roll and pitch orientation, so that both must be actively stabilized.[4]

- Thrust is mainly provided through the main rotor. Horizontal force components can only be exerted by tilting the whole vehicle.

- There are no fixed control surfaces available. Control needs to be effected by varying the pitch angles of the rotor blades.

Figure 1.5 visualizes the flight controls setup met in a helicopter. In order to obtain as fast as possible responses, the RPM values of both rotors are always kept as constant as possible in helicopters, with the exception of very small toy vehicles. Force and torque control is performed through blade pitch changes only. For the *tail rotor pitch*, this is a single value determining the yaw torque and controlling the helicopter's yaw rotation. In the case of the main rotor, things are more complicated, for it must produce both a variable lifting force and variable roll and pitch torques. For this purpose, a clever device in the rotor head, the *swash plate*, changes the main rotor blades' pitch values during their rotation, mechanically establishing

---

[4]Actually, all helicopters show a certain tendency of passive stabilization relative to the surrounding air, which is basically due to the fact that a rotor blade produces more lift when moving against than when moving with the relative horizontal airflow. Recently, some in-door toy helicopters have been introduced that fly perfectly stable without any active control of the main rotor pitch. However, this effect is too small – and does not suitably scale with vehicle size – to rely on in larger size, faster moving helicopters.

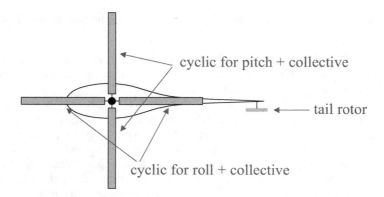

Figure 1.5: Flight controls outline of a helicopter-type vehicle.

Control	Type	Effect
collective pitch	constant blade pitch offset	lift force
cyclic pitch – pitch	cyclic blade pitch offset	roll torque, pitch rotation
cyclic pitch – roll	cyclic blade pitch offset	pitch torque, roll rotation
tail rotor pitch	constant blade pitch offset	yaw torque

Table 1.3: Summary of typical helicopter controls.

a function of rotation angle to pitch angle. Therefore, figure 1.5 shows a four-blade rotor to indicate the blade positions that correspond to pitch and roll torques. During the rotation of the main rotor, these positions remain unchanged, but the actual blades take turn in exerting the respective functions.

Roll and pitch torques applied to a helicopter in flight primarily change the angular momentum of the main rotor – the angular momentum of the fuselage's rotation can safely be neglected (see section 5.5.2 and [38] for a more detailed treatment). Assuming counter-clockwise rotation of the main rotor, i.e. an angular momentum vector directed upward, a torque vector in forward direction will also tilt the main rotor's angular momentum forward, resulting in a forward pitch of the fuselage. This is just an applied formulation of the law of conservation of angular momentum. However, said torque is actually a "roll" torque, i.e. generated through maximum blade pitch on the left and right sides of the fuselage. This is why the left and right blade positions are depicted in the figure to be responsible for pitching, whereas the front and aft positions are responsible for rolling. These settings are called *cyclic pitch*, due to their change with the rotation of the blades.

Additionally, the regulation of the lifting force is accomplished through a blade pitch offset that is constant throughout the rotor circle, called *collective pitch*. The effect of collective pitch alone corresponds to the pitch control mechanism found at the tail rotor. All in all, the swash plate administers three degrees of freedom by way of the main rotor, while the tail rotor takes care of one degree.

Table 1.3 summarizes the available helicopter controls explained above and their effects.

1.4 Overview

This final section of the introductory chapter provides an overview of the remaining chapters of the book.

The following chapter 2 presents a non-representative collection of example systems in some greater detail. This will already provide the reader with some insight into typical characteristics of small UAV systems, without further elaborating them formally. Naturally, the creators of most of these example systems have already been mentioned in the preceding section 1.2.

The subsequent chapters deal with one of the "typical" task areas each:

Chapter 3 covers the selection of on-board computing equipment and – possibly – operating systems. Chapter 4 discusses the aspect of intra-system communications. This concerns communication devices suitable for on-board use on the one hand, and middleware for providing abstract views of the cooperation of distributed system components on the other.

Chapter 5 presents the requirements to small UAV flight control and introduces a novel design approach suitable for UAV controllers that has been developed in the author's group. Some examples of controller architecture for different vehicle types are presented in order to illustrate the suitability of this design approach.

Chapter 6 deals with sensor classes that are typically used on board small UAVs, both for flight control and for mission data acquisition purposes. Common for all autonomous aerial vehicles is the requirement to measure position and orientation as inputs to the flight controller. The chapter will also deal with fusion algorithms to transform raw sensor measurements into this kind of more abstract data.

Chapter 7 finally discusses the development process connected with the practical design of a small UAV and tools and methods that tend to proof useful (i.e. time-saving) during the former.

Chapter 8 concludes the book and summarizes its findings.

Chapter 2

Example Systems

This chapter presents a selection of example small UAV systems that have been developed in the last couple of years and that have been proven to be operational – to some degree at least. Not surprisingly, all of the research groups involved in the development of these systems have already been mentioned above in section 1.2.

Following the most intuitive classification according to the type of aerial vehicle used, the chapter is partitioned into airship UAVs (section 2.1), fixed-wing plane UAVs (section 2.2), and helicopter UAVs (section 2.3).

2.1 Airships

Airships constitute the most easily manageable type of UAV, because they are basically able to fly without any kind of active control and without any moving parts and energy supply. But flying as such is one thing, controlled motion is another. The main disadvantages connected with airship UAVs are:

- Their *size* leads to adverse aerodynamical properties, resulting in slow motion and extreme susceptibility to wind.

- The need for helium[1], which in turn is difficult to keep inside the hull permanently, often causes high operation costs.

- With variations in environment temperature and solar radiation, it may be difficult to sustain lift trim without dropping ballast weight or discharging helium.

However, inherent safety and simplicity have lead to the development of several airship UAVs. The ones covered here are *TubRob* by TU Berlin (section 2.1.1) and *Karma* by LAAS/CNRS (section 2.1.2).

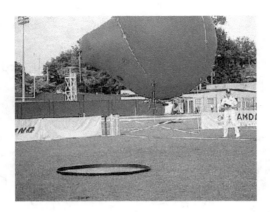

Figure 2.1: Airship UAV *TubRob* by TU Berlin during the 1995 IARC.

2.1.1   TubRob

TubRob is an ancient non-rigid airship ("blimp") type UAV developed between 1993 and 1995 by the Real-Time Systems Group (PDV) at TU Berlin. Its purpose was to participate in the 1995 International Aerial Robotics Competition (IARC) [46], where it finally took 2nd place [23, 22].

TubRob's key specifications comprise:

- spherical balloon hull of about 2.5 m diameter, filled with helium,

- lift for approximately 7 kg total mass,

- 6 electrically powered propellers for propulsion, sideward thrust, and lift regulation, self-manufactured power regulators,

- maximum airspeed approximately 2 m/s,

- carbon frame with plastic net serving as ultra-low-weight instrument platform,

- 2 on-board microcontrollers SAB80C166 [12],

- i486 PC plus one SAB80C166 microcontroller constituting the ground station,

- communication with ground station via 2400 bit/s RS 232 interface modulated onto one HF radio link per direction.

As TubRob was designed before affordable GPS receivers with suitable performance became available, the position sensor system had to be specifically developed by the TubRob team. This system, nicknamed *global balloon positioning system* (GBPS), consists of a 360° circular

---

[1] As everyone knows, the use of hydrogen is even less advisable.

ultrasonic transmitter made from 12 transducers by Polaroid (see section 6.2.5 for a roughly similar device) and 6 receivers at the border of the operation area, using the same kind of transducers. The periodic transmission of ultrasonic pulses is synchronized via the frame sync signal of an on-board video camera with a PLL (phase-locked loop) algorithm running on the ground station microcontroller, which also reads the reception signals from the ultrasonic receivers. With the operation area measuring only 20 m × 40 m, this system works reliably and provides a positioning accuracy of a couple of centimeters at worst.

For altitude measurement, additional ultrasonic sonar sensors are provided. Using simple piezo transducers, their operation range is limited to about 2 m maximum altitude.

Orientation sensing is only required for yaw, because roll and pitch angles are passively stabilized through the weight of the instrument platform. The yaw angle is measured by a single-angle compass sensor of type 1525 by Dinsmore Sensing Systems [133]. This compass uses a rotating magnet in place of a compass needle and detects this magnet's orientation by means of two hall sensors.

TubRob's flight controller calculates desired RPM values for the six propellers based on an RPM-thrust-model. These RPMs are instantiated through RPM controllers and photo sensors at each propeller. The position controller's operating point is used to estimate the direction of atmospheric wind in order to turn the UAV into the wind since the 2 propulsion motors are more powerful than the remaining 4 navigation motors.

As the biggest threat to an airship UAV consists in loosing it to the skies in consequence of an undesired too-positive weight trim, TubRob is equipped with a clever safety system: An electric magnet holds a weight connected to a long thin safety line fastened to the UAV. The safety line drops to the ground when the magnet is switched off, on-board power is lost, or a simple mono-flop watchdog circuitry is not periodically reset by the microcontroller.

Figure 2.1 shows TubRob during the 1995 competition in Atlanta, Georgia. While the competition's primary task, the grasping of specialized double metal disks of 10 cm diameter, had been successfully demonstrated in indoor experiments with the TubRob system, its accomplishment during the competition was hindered by too much wind – although still moderate from an every-day perspective. Hence, the first place in the competition went to the team from Stanford University, who used a helicopter UAV and thus demonstrated that helicopters are preferable performance-wise.

2.1.2 Karma

Karma is an airship UAV developed by LAAS/CNRS in Toulouse [25, 89]. It has been primarily designed for stereo image acquisition and participated in the COMETS EC project [28].

Key specifications of this UAV include:

- based on AS-500 by Airspeed Airships,
- length 7.8 m, maximum diameter 1.8 m, volume ≈ 15 m^3,

Figure 2.2: Airship UAV *Karma* by LAAS/CNRS, Toulouse, during the 2005 COMETS experiments.

- maximum payload 5 kg,

- 4 control rudders in X shape,

- electrically powered, tiltable main thrusters,

- EBX form factor PC motherboard with Celeron 566 MHz CPU.

Karma is equipped with the following on-board sensor systems:

- Precision Navigation TCM2 compass and "inclinometer",

- Trimble Lassen SK II GPS receiver,

- Vitana PL-A633 stereo cameras with IEEE 1394 interface,

- LCJ Capteurs CV3F wind and airspeed sensor.

The TCM2 compass module [34] combines a 3-axis magnetometer and a 2-axis acceleration sensor. It is generally only suitable in static situations, as any motion acceleration interferes with its tilt measurement (see section 6.1 for a discussion). Yet, operation on board an airship is a good enough approximation to static usage, with its motion acceleration being rather limited.

Figure 2.2 shows Karma during the 2005 COMETS experiments in Lousã, Portugal. Until then, fully autonomous operation had not been implemented in Karma. However, its main purposes, like image acquisition and map generation, were successfully demonstrated under remote control.

2.2 Fixed-Wing UAVs

Concerning design complexity, cruising range, and miniaturization, fixed-wing UAVs clearly constitute the optimum choice of air vehicle type. Aerodynamically stable fixed-wing planes can be successfully navigated with 2 control inputs only (e.g. thrust plus rudder, or two independent thrusters), which is not possible for any other usable construction.

The main drawback of plane-type UAVs consists in their inability to hover, resulting from the stall speed's imposing a lower bound on the vehicle's airspeed. However, when considering application scenarios, it seems that this drawback is more relevant to the design phase than at any later occasion. Of course, in all flight experiments the UAV must be prevented from getting out of visual and/or radio contact. However, this can be achieved without significant difficulty, by manual remote control as a safety procedure (see section 7.3.1) on the one hand and the implementation of automatic circle flight, which can effectively substitute hovering, on the other.

Numerous small fixed-wing UAVs have already been developed, both in experimental fashion and as commercial products. Therefore, this book only selects two very arbitrary examples, the *Black Widow* micro air vehicle (MAV) developed by AeroVironment (section 2.2.1) and *Carolo* by TU Braunschweig (section 2.2.2).

2.2.1 Black Widow

The *Black Widow* is the most legendary offspring of the DARPA MAV program initiated in 1996. [61] provides a quite comprehensive and detailed report of its development and features, especially in view of its clearly military background.

At the beginning of its development, the 15 cm size restriction imposed by DARPA constituted the primary challenge to all companies and groups participating. AeroVironment Inc. of Monrovia, CA performed numeric optimization, in part using genetic algorithms, in order to determine the optimal combination of components (aerodynamics, batteries, propeller, motor, gearbox) with respect to the mission time possible.

Due to its size, it is inaudible and almost invisible even at moderate flight altitude (e.g. 30 m, according to [61]). The UAV's reported key features:

- wingspan 152 mm (6 inches),
- mass 80 g,
- endurance 30 min,
- cruising speed $\approx$ 11 m/s,
- control surfaces rudder and elevator, moved by 0.5 g actuators,
- RF digital command uplink at 433 MHz,
- RF analog video downlink at 2.4 GHz.

Avionics sensors on board are provided for airspeed, compass heading, and yaw rotation rate. Semi-autonomous operation has been implemented through three alternative autopilot modes, which are airspeed hold, altitude hold, and heading hold. For mission data acquisition, the "final" version features a color CMOS camera weighing 1.7 g and a video transmitter with 100 mW RF output weighing 1.4 g only. The video feed can be used for out-of-sight human remote control, which is the primary mode of operation.

On August 10th, 2000, AeroVironment performed a demonstration flight with the Black Widow that finally fulfilled the 30 min endurance objective. This flight exhibited a maximum range of 1.8 km and a maximum altitude of 234 m.

Interestingly, AeroVironment has *not* included a successor of the Black Widow prototype system, or any UAV below 40 cm wingspan, in their product list as of 2006. This indicates that efforts in the 15 cm class are still to be considered feasibility studies, while commercial gain is not yet being anticipated. This probably results from the very limited payload and the limited level of autonomy in today's incarnations of this class. Yet, further miniaturization and cost reduction of commercially available sensor, communication, and processing devices may soon open commercial opportunities here.

2.2.2 Carolo

The Institute of Aerospace Systems (ILR) at TU Braunschweig has been developing "mini" UAVs for several years, with the first flight of the fixed-wing UAV prototype *Carolo* conducted in 2002 [87, 24]. The ILR's declared objective is to implement fully autonomous control, which is why they have not yet targeted the "micro" UAV class as originally specified by DARPA. Meanwhile, a big portion of UAV development in Braunschweig has shifted from the ILR and its scientific nature to the spin-off company Mavionics GmbH, founded in 2004.

The smallest UAV system of the Carolo family having approached product maturity is the *Carolo P50* [24, 142]. It flew autonomously for the first time on April 7th, 2004. Its key specifications include:

- wingspan 49 cm,

- maximum take-off mass 450 g,

- maximum payload mass 50 g,

- endurance $\leq$ 15 min,

- range 1000 m with telemetry data only, or 500 m including video transmission,

- optimal cruising speed $\approx$ 15.5 m/s.

The P50 is equipped with a full avionics unit, consisting of a full 6-DOF inertial measurement unit (IMU), a GPS receiver, and a 32 bit/200 MHz microprocessor. This permits fully autonomous flight control with arbitrarily sophisticated algorithms. Special payload sensors are provided with the focus on meteorological applications.

2.3 Helicopters

Helicopter UAVs combine optimum maneuverability, hovering capability, minimum required space for take-off and landing, a good payload-to-total mass ratio, and sufficient propulsion reserve for operation in windy conditions. Therefore, they constitute the prime choice in many application scenarios. On the other hand, there are also significant disadvantages, especially their instable flight behavior, mechanical complexity, risks due to the great amount of stored rotational energy, and adverse acoustic signature.

While the remaining disadvantages may be considered gradual, the instability of flight behavior strictly requires, in contrast to the UAV classes discussed so far:

- an attitude control system that permanently stabilizes the UAV's pitch and roll angles, and

- pitch and roll angle sensors that are reliable and accurate enough for this purpose.

Examples selected for this class are the Ursa Electra helicopter of the Berkeley Aerobot Team (section 2.3.1), the ARTIS demonstrator by DLR (section 2.3.2), and the MARVIN UAV (section 2.3.3), which has been developed in the author's group at TU Berlin. Finally, TU Berlin's quadrotor testbed system for indoor flight is presented in this section, since quadrotors are similar to helicopters in their flight dynamics and maneuverability.

2.3.1 BEAR Ursa Electra

The *Berkeley Aerobot Team* (BEAR) operates several autonomous helicopter UAVs, some of which clearly exceed the limit of a "small" UAV as imposed by this book. However, BEAR's *Ursa Electra* helicopter is one of the first electrically powered helicopter UAVs capable of autonomous outdoor flight with usable endurance and, hence, constitutes a suitable example system.

Ursa Electra's key features include [118]:

- based on Maxi-Joker small-scale helicopter by Joker USA,

- rotor diameter 1.8 m,

- basic mass 4.5 kg,

- maximum payload 4 kg,

- endurance 15 min,

- on-board CPU Pentium 3, 700 MHz,

- energy source Lithium-Polymer batteries.

It carries a full set of avionics sensors, namely an OEM4 GPS receiver by NovAtel and an Inertial Science ISIS inertial measurement unit plus an additional (probably magnetic-field-based) yaw angle sensor. This vehicle provides good evidence that today, the benefits of an electric helicopter need not be obtained at the expense of too much operation time any more.

2.3.2   ARTIS 1.9 m Helicopter

Within the ARTIS (Autonomous Rotorcraft Testbed for Intelligent Systems) project of DLR, Germany, several helicopter UAVs are under development. This book is to focus on the initial one of these, the ARTIS helicopter with 1.9 m rotor.

The design of this UAV has not been optimized for weight or performance, but great efforts have been made to modularize the on-board system as far as possible. Thus, different components have been built into separate housings, which in turn can be combined in a rail arrangement mounted to the helicopter. This is optimally suited for fast and easy replacement of individual components, at the expense of an otherwise avoidable amount of total housing and wiring mass to be lifted.

Due to this modularity, a universal description of this UAV is more difficult than with other systems. Nevertheless, general features include [39, 134]:

- rotor diameter 1.9 m,
- payload > 6 kg (including avionics etc.),
- PC-family on-board computer,
- GPS receiver,
- inertial measurement unit and compass,
- sonar altitude sensor,
- wireless LAN and radio modem for communication,
- equippable with various camera systems.

This UAV has been flying autonomously in a routinely fashion since 2004. Research conducted by DLR that involves the ARTIS demonstrator includes dynamic collision avoidance and visual position control.

2.3.3   MARVIN Mark II

The Real-Time Systems Group (PDV) at TU Berlin began the development of helicopter-based UAVs in 1997. The first UAV resulting from this development, MARVIN "Mark I", won the IARC three-years-final in 2000 [99]. Keeping all basic design approaches, further development in the context of the EC COMETS project lead to the current development stage,

Figure 2.3: MARVIN Mark II UAV as of 2005 (left), rotor head and upper gearbox detail (right).

MARVIN "Mark II", which will be described in some detail here. With the author's having been a member of the MARVIN team from the start, certain aspects of this system will be examined in still greater depth later throughout this book. In this section, "MARVIN" always refers to the "Mark II" stage. The design of this stage was coordinated by Volker Remuß [114, 68], who also provided most of the visual material used here.

Figure 2.3 (left part) gives an overview of the MARVIN UAV while standing on the ground. MARVIN is based on a commercially available professional RC helicopter, the CB-5000 by Aero-Tec [1]. In contrast to the majority of model helicopters designed "for fun", it features a solid reliable construction with, e.g., metal gears and an all-metal rotor head originally developed by Dieter Schlüter. A detailed view of this rotor head and the upper gearbox is provided in the right part of figure 2.3.

Key features of the base helicopter are:

- basic mass (no avionics, no camera) ≈ 8 kg,

- rotor diameter 1.84 m,

- 23 cm^3 2-stroke petrol engine, 1 cylinder, 1.8 kW,

- electrical engine starter.

All components necessary for remote-controlled flight reside in the "off-the-shelf" upper part of the helicopter. The black front part contains the RC receiver and batteries for starter, receiver, and servos. Almost all components for autonomy and image acquisition are combined in the black carbon-fiber avionics box under the red landing gear. Figure 2.4 depicts this box with the top-lid removed and indicates the location of the individual components:

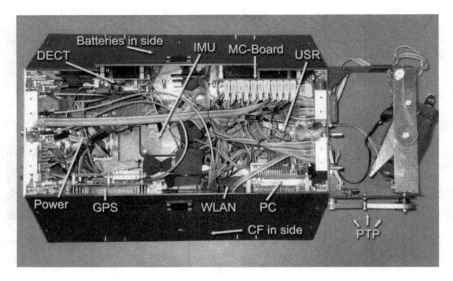

Figure 2.4: MARVIN's avionics box with explanations.

- a DECT radio modem for digital data communication (see section 4.2.4),

- an inertial measurement unit (IMU) of type 3DM-GX1 by Microstrain (see section 6.2.6),

- a circuitry board carrying the on-board microcontroller, an SAB80C167 by Infineon (see section 3.2.3), and most of the peripheral device connectors,

- an ultrasonic altitude sensor module (USR, see section 6.2.5),

- a custom-made power supply,

- an OEM4 GPS receiver by NovAtel (see section 6.2.1),

- a system-on-a-chip PC by Compulab, Israel, equipped with a Geode SC2200 at 266 MHz, running Microsoft Windows XP Embedded from a CF memory card,

- a wireless LAN PC-Card in the on-board PC serving as high-bandwidth data downlink and communication backup,

- a custom-build carbon-fiber pan-tilt-platform (PTP) carrying a Canon PowerShot S45 digital photo camera.

In this setup, the PC is not required for autonomous operation. Instead, its main purpose is to control the on-board camera via a USB connection and send the images to the ground station via WLAN. This is why a non-real-time-capable operating system can safely be used on the PC.

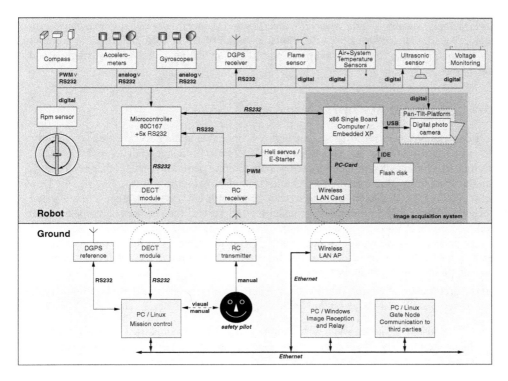

Figure 2.5: MARVIN's system architecture.

Together with the avionics box and camera, the total take-off mass of MARVIN amounts to 12.7 kg, including 650 cm^3 of fuel. This enables the UAV to fly autonomously for about 30 min. Thanks to the electric engine starter, the autonomy phase may start and end with the engine at rest, while everything from starting the engine, take-off, flight, landing, and stopping the engine again is performed under the control of the on-board computer.

Figure 2.5 summarizes the full architecture of the MARVIN system. Beyond the on-board components already discussed, the ground station shall be addressed with the help of this figure. It consists of one or more PCs with the necessary communication hardware and the GPS reference receiver connected to them. The most important "component" of the ground station, however, is the human safety pilot (see section 7.3.1). He or she carries a conventional remote control transmitter, which can switch the UAV between manual remote control on the one hand and autonomous operation on the other.

Particularly notable to UAV designers should be the fact that the RC receiver on board is connected to the on-board computer via a serial (RS 232) interface. This is a special feature of "diversity" RC receivers manufactured by ACT, Germany [140]. The ACT DDS10, which is used aboard MARVIN, outputs the RC input signals received via a serial protocol and can be instructed to output servo controls provided via this interface instead of the received ones.

This is, actually, an abuse of the receiver's diversity and trainer modes rather than a purposeful feature. Nevertheless, it saves both the implementation of about 16 digital input and output lines to and from the on-board computer and of the switching logic between manual and autonomous modes.

2.3.4 Quadrotor Testbed

Quadrotor vehicles have recently become highly fashionable in the UAV research community. This is first due to the fact that small quadrotor UAVs can be designed in a very simple way, with four RPM-controlled propellers as their only moving parts, and on the other hand results from the availability of attractive base vehicles from the toy industry of late, like e.g. the DraganFlyer V Ti [73]. However, fully autonomous quadrotor UAV systems are just on the verge of becoming available. Therefore, this chapter briefly presents the new indoor testbed system developed in the Real-Time Systems group of TU Berlin, under the direction of Konstantin Kondak and Markus Bernhard, as an example. The main purposes of this system are the investigation of collision-avoidance sensors and strategies, and the design of future quadrotor UAVs of similar size for outdoor usage.

This quadrotor uses, of course, four fixed propellers directly driven by external rotor brushless motors. Being an indoor-only system is reflected in two essential design characteristics:

- The position sensor consists of a number of external cameras tracking light bulbs attached to the UAV. This is an effective approach given that GPS cannot be used indoors.

- The UAV does not carry batteries, but power is supplied by cable. This increases the available payload and avoids the troubles of recharge cycles and battery wear.

Flight control is performed by a ground-based PC running Windows CE and a MATLAB Embedded Target controller. Control inputs and sensor readings are transmitted to and from the vehicle via an RS 232 interface cable. On board, an Infineon SAB80C167 microcontroller is responsible of hardware interfacing. The only on-board sensors strictly required for autonomous flight are 3 rotation rate sensors connected to the microcontroller. For RPM control, 4 off-the-shelf RC brushless controllers are used, but with RPM control mode disabled.

The technical features of the quadrotor UAV include:

- propeller diameter 39 cm,

- total take-off mass approx. 5 kg,

- frame size 0.8 m $\times$ 0.8 m $\times$ 0.2 m ,

- power consumption in hover approx. 800 W,

- propellers fully inside the vehicle's aluminum frame for safety reasons.

The testbed system has successfully demonstrated its capability of fully autonomous and highly dynamic indoor flight. Figure 2.6 shows the vehicle in flight, with the positioning bulbs clearly visibly.

Figure 2.6: Indoor quadrotor testbed system at TU Berlin in flight.

2.4 Subsumption

After reviewing UAV systems available throughout the world, it is obvious that designing a small UAV system is possible today with moderate effort. Only extreme miniaturization, like e.g. in the DARPA MAV initiative and the Black Widow UAV resulting from it, complicates the design process, leads to extreme costs, and minimizes or possible removes the margin for potential applications. Considering the possible vehicle types to base UAVs on, the appropriate choice depends on the requirements – every option may be reasonable in some particular application (or research) scenario.

Chapter 3

On-Board Computers

In order to perform flight control and to interface a UAV with some manned or unmanned ground station and, possibly, other UAVs that may belong to the overall system as well, it is quite clear that every UAV needs to carry one or more processing units. This chapter illuminates the developer's options in selecting the type, model, and number of processing units and in the choice of an operating system.

Section 3.1 starts this chapter with an overview of the requirements to on-board processing units. This way of introduction will be common among all the remaining thematic chapters, in order to obtain a clean separation of the "what" and "how" issues related to each of the aspects of small UAV system architecture as covered within this book. The following section 3.2 deals with the issues of CPU selection, divided into questions of architecture and those of actual product families. The final section of this chapter, section 3.3, elaborates the decision on – and for – an operating system to provide basic services to the software modules that take care of the UAV's operation. Some current operating systems with suitable real-time capabilities are specifically addressed, without going into too much detail, though. The final section 3.4 summarizes this chapter's findings in compact form.

3.1 Requirements

This section lists the requirements to be typically met by the on-board processing unit or units of a small UAV. These requirements can be divided into the task and data processing needs, the possibilities of interfacing peripherals, and the issue of development support available for embedded processors.

3.1.1 Task and Data Processing

Figure 3.1 suggests a classification of tasks to be performed by small UAV processing units. This is, clearly, not the only possible classification, but it should suffice to fit any actual system as well as to illustrate the basic data processing paths. The latter serve as the basic distinctive criteria here.

UAV "application" software

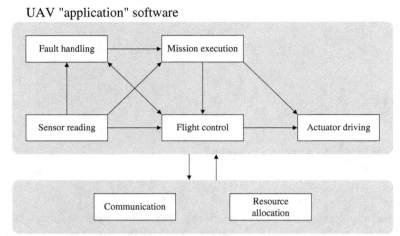

Operating system and related software

Figure 3.1: Overview of basic on-board computing tasks, their interaction, and possible division into "application" and "operating system" aspects.

First, figure 3.1 groups the tasks to be performed into

- "application level" tasks, which are specifically concerned with the operation of the UAV, and

- "operating system level" tasks, which are concerned with the distributedness of the system and the operation of the processing units.

On the application level, there are the following classes of tasks:

Sensor data processing tasks are concerned with the interfacing of sensor components, the preprocessing of sensor data, and the fusing of data from different sensors.

Actuator driving tasks are concerned with the passing of control signals to the actual hardware components that act as the system's actuators.

Flight control is the software implementation of the flight controller. Functions grouped under flight control are purely concerned with the UAV's motion, i.e. the trajectory of its position and attitude.

Mission execution tasks are concerned with application-specific aspects of the UAV's behavior. This may on the one hand refer to the employment of some mission-specific sensors and actuators, on the other to mission-specific criteria on flight control's mode of operation.

Fault handling tasks are concerned with the detection, identification, and handling of occurrences of faulty or non-optimal behavior of any component of the UAV. This includes sensor failures or the deterioration of sensor data quality as well as actuator failures. While fault handling tends to be complex and requires the individual addressing of many situations, it is of course highly desirable to prevent the most common classes of failures from causing total loss of control.

On the operating system level, the most typically encountered tasks are:

Communication tasks, being concerned with data exchange between different components and/or processes within the UAV system. From the programmer's point of view, communication services can be regarded as providing an abstraction from the distributedness of the system. Software that offers this kind of abstraction is usually referred to as *middleware* (e.g. [4]).

Resource allocation tasks, being the most "typical" operating system tasks, such as memory management, process scheduling, and storage device management.

These "operating system level" tasks are considered to interact with all the "application level" tasks mentioned above.

As all of the tasks discussed in this section depend on certain external events *and* may affect the UAV's motion, the operating environment of the task software – possibly in the form of a universal operating system – needs to guarantee upper bounds on all kinds of response times. Therefore, the operating environment of the task software – possibly in the form of a universal operating system – needs to be capable of real-time operation (e.g. [129]). This clearly prohibits the use of any typical non-real-time, "desktop" type operating system, e.g. from the Windows and Unix families.

3.1.2 Peripheral Interfacing

Small UAVs are classical *embedded systems*, requiring on-board processing units and a variety of sensors and actuators to act together. As a consequence, the physical interfacing between processors and *peripheral devices*, i.e. sensors and actuators, is one of the central tasks in UAV design. The crucial aspect here is the fact that every interface concerns, at least, two components. Changing some processing unit to another type that does not support an interface type that has been in use beforehand, as well as replacing some sensor with another one carrying an interface type unsupported by the on-board processing units, usually results in a major system redesign, replacing several on-board components at once.

Below, this section discusses the interface types common among sensors, actuators, and processors suitable for use on board small UAVs.

Digital IO Lines

Digital, i.e. binary state input/output lines are both very simple and widespread, but limited to fairly simple interfacing tasks. Every digital IO line can represent one of two logic states, usually referred to as *HI* and *LO* or as 0 and 1, expressed through intervals of voltage level for each of the two states. This is, of course, just the way any digital signal inside an integrated circuit or circuit board works. In order to express more than two states, n digital IO lines can be grouped, resulting in 2^n discernible states.

As the number of available input and output lines tends to be one of the most limited resources of the processing components, simple digital IO is typically used only for dedicated peripherals with a very limited number of states to be communicated. Possible examples of the use of digital IO lines include: ground contact sensors (switches), limit switches of manipulators or movable camera heads (pan-tilt-units), engine cut-off signals, low-fuel sensors, arbitration between automatic and manual control, and indicator LEDs.

Analog IO Lines

Analog input/output lines are conceptually simple and basically capable of carrying a great amount of information per line. The information to be transmitted is represented as an "analog" voltage level within a permissible interval of voltages. While many processing units suitable for embedded UAV systems are equipped with *analog-to-digital converters* (ADCs), meaning they are capable of reading analog inputs, *digital-to-analog converters* (DACs) are comparatively unusual. Therefore, considering analog IO lines can, in most cases, be restricted to the notion of analog *inputs*.

When comparing analog interfacing capabilities, the converter's *resolution*, in bit, is of crucial importance. It determines how many different voltage levels can be distinguished when the conversion to or from the digital (numeric) representation for use within the processing unit is performed. An ADC with 10 bit of resolution can, for example, distinguish $2^{10} = 1024$ different values. Usual resolutions in current integrated processing units tend to be in the range between 10 bit and 16 bit. Depending on the application and the sensor type in question, the available converting resolution may constitute a serious limitation.

Another performance criterion on analog-to-digital converters is the maximum sampling rate available. This is due to the fact that the conversion is often performed as a sequential process by the ADC's circuitry. Additionally, it is quite common to have a small number of actual converters, but a significantly higher number of analog input lines available. Then, subsets of the latter are *multiplexed* for sequential conversion by a single ADC in each case. Consequently, the maximum available sampling rate per analog input depends both on the maximum conversion rate of the ADC and on the number of analog input lines to be sequentially processed by it.

Among the typical attitude sensors for use in small UAVs, analog outputs have been quite common for some time. Currently, a certain tendency can be observed of replacing analog outputs with digital PWM and bus-type interfaces (see below). This is probably due to certain typical disadvantages of analog interfacing:

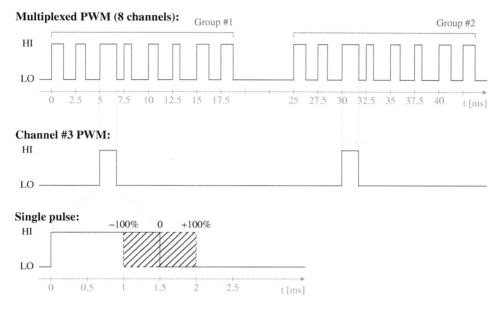

Figure 3.2: Timing scheme of remote control pulse-width modulated signals, channel-multiplexed and single channel ("servo signal").

- Frequently, a separate, highly stable reference voltage is required by the converter circuits.

- Analog signal lines are highly susceptible to interference from electrical noise, because any subtle change in their voltage will be misinterpreted as information.

- Analog outputs produced by sensors rarely have an amplitude suitable for the processing components in use. Normally, operational amplifiers must be added to convert the respective voltage ranges. Frequently, the analog interfacing circuitry also needs to include filters, in order to prevent high-amplitude noise from further reducing the resolution effectively available. Further more, such kind of analog interfacing circuitry may require special supply voltage levels not usually present in computing circuitry, e.g. "negative" voltage.

Useful applications of analog inputs include: Battery voltage monitoring, magnetic field, acceleration, rotation, and airspeed sensors, temperature monitoring (internal or environmental).

Pulse-Width Modulated Signals

Modulating a value to be transmitted as a period (or pulse-width) of an oscillating digital signal, called *pulse-width modulation* (PWM), combines the advantages of digital and analog IO lines: Only one line is required for each value, and the digital signal level is not significantly susceptible to analog noise. Due to the high slew rate of the digital signal, analog noise can only have very limited effect on the pulse width as detected by the receiving circuit. Finally, due to the high CPU clock rates, the resolution of pulse-width measurement usually found in processing units for embedded systems is typically superior to the resolution achievable through analog-to-digital converters. Of course, the maximum resolution of the receiver's pulse width detection circuit imposes an upper bound on the *product* of sampling rate and sampling resolution.

Pulse-width modulation can be discriminated into two different flavors, namely coding via the *frequency* or via the *duty-cycle* only. In the former case, the length of a full signal period is modulated, which effectively results in changes of the sampling rate together with the value. In the latter case, the signal period – and such the sampling rate – remain constant, while the value is coded through the relative duration of the HI and LO periods of the signal, respectively.

A particularly prominent application of PWM coding in the small UAV sector consists in the use of remote control (RC) components as common in model making. These components are easily and affordably available for the full chain of radio frequency transmitters, radio frequency receivers, and actuators like servos and motor regulators. While there does not seem to be any "official" standard describing the communication between these components, all common systems do use PWM coding at least for single-actuator control. Figure 3.2 depicts this common coding scheme considering an hypothetic 8-channel remote control as an example: The signal coding all 8 channels, which has to be transmitted via radio-frequency modulation, consists of groups of 8 pulses, each of which refers to a single channel. The pulses start with a positive transition each and are 2.5 ms apart. Each group is terminated by means of a gap of 5 ms. The remote control receiver would obtain this signal from demodulating the received radio-frequency signal[1] and perform sequential time-demultiplexing to obtain 8 single-channel signals like the one depicted in figure 3.2 for the third channel. These signals are used for servo control. The bottom part of figure 3.2 presents a magnified view of one of the pulses: Its HI-level duration codes the channel's state, with 1.5 ms coding zero (centered), 1.0 ms coding -100 %, and 2.0 ms coding $+100$ %. The duration of 1.5 ms for "zero" pertains to almost all commercially available remote control systems, so it can be considered a de facto standard. However, the pulse rate (which is $\frac{1}{25\,\text{ms}} = 40$ Hz in the example) may vary depending on the system and even its settings.

With the knowledge of this PWM protocol, UAV designers can easily employ RC transmitters to send manual commands to the on-board computers, as well as generate the required control signals to drive conventional servos from the on-board computers.

Apart from RC components, all sensor types mentioned above for analog interfacing are available with PWM outputs as well. That is: voltage, temperature, and attitude-class sensors.

---

[1] In modern RC systems, there are more fault-tolerant alternatives available for transmitting the multiplexed signal, like digital coding (also called PCM for *pulse-code modulation*). However, the single-channel output signals of the receiver are still PWM-coded in this case, just as explained above.

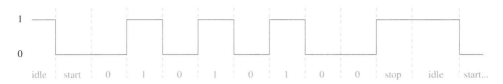

Figure 3.3: Timing scheme of RS 232 asynchronous serial interface, transmitting one byte with decimal value 42.

Asynchronous Serial Interface (RS 232 C)

The notorious "serial port" is one of the most-widespread and best-known digital interfaces for connecting two different devices. It is based on two corresponding standardization documents [18, 136], which initially go back as far as the 1960s. The original purpose of the interface was the connection between a teletypewriter terminal and a modem for data transmission.

RS 232 is certainly the simplest available interface for byte-oriented data transmission. It requires only one data line per direction (transmission versus reception) plus one line for ground, in a minimal configuration. This simplicity results from the asynchronous nature of the signal timing, as shown in figure 3.3 for the transmission of one byte with value $42_{10} = 00101010_2$: initially, the data line is at "1" level, representing *idle* state. Then, the 1-0-edge of the *start* bit ("0" level) initiates the time frame for the following byte. This is possible because the bit duration is known by both peers by prior arrangement on the bit rate. After the start bit, the eight data bits of the byte are sent, starting with the least significant bit. Finally, the sender adds one *stop* bit to ensure the line's being at "1" level again after the transmission. The subsequent idle phase may be of arbitrary duration or completely missing, for the synchronization for the next data byte will be determined by the subsequent *start* bit. Thanks to this simple structure, it is even possible, worst comes worst, to do first-glance analysis of RS 232 communication using just a standard oscilloscope.

RS 232 data output is very common with "intelligent" sensors, i.e. sensors equipped with some microcontroller for post-processing of data. Furthermore, devices with RS 232 interfaces to connect sensors of this kind to are widespread, in the form of traditional visual terminal devices or desktop PCs. This kind of setup is very helpful especially during development. Embedded-style processing units almost always support RS 232, or at least allow the adding of such interfaces as peripheral devices. However, it has to be noted that RS 232 support is degrading today. Modern notebook PCs, and more and more desktop systems as well, lack native RS 232 interfaces, mainly in favor of the more current USB technology (see below). But this transformation does not yet fully project to the sensor market; so in laboratory and embedded system environments, RS 232 is still being applied very frequently. For PC interfacing, USB-to-serial converters may be applied to reestablish the "missing" ports.

There are some disadvantages of RS 232: The standard determines voltage levels of up to ± 25 V, which have to be specially generated by dedicated driver circuits (*MAX232* chips and derivatives). Some, but not all, devices with RS 232 interface also support $0\,\text{V} - 5\,\text{V}$, but others do not, so the interface voltage level is always a major design concern with RS 232.

Feature	Value(s)
Data rate	$\leq$ 480 Mbit/s
Power supply via bus	$\leq$ 500 mA @ 5 V
Transmission modes	interrupt, bulk, isochronous
Reconfiguration	dynamic (*hot-swappable*)
Single cable length	$\leq$ 5 m
Host-to-device distance	$\leq$ 30 m
Number of connected devices	$\leq$ 127

Table 3.1: List of USB features and capabilities according to the USB 2.0 specification.

Furthermore, the connectors defined by the standard documents contain a great number of dedicated control lines that are seldom required in embedded system applications, so for the latter, it can be stated that there is no suitable connector standard at all. Finally, the data rates supported by RS-232-capable devices and circuitry tend to be quite limited, mainly for historical reasons. The PC architecture has never safely supported more then 115200 bit/s, and there is a special list of "usual" data rates (300, 600, 1200, 2400, 4800, 9600, 19200, 38400, 57600, 115200 bit/s) that can as well only be explained historically. Support for higher rates is technically possible without any problem but seldom found.

Typical applications of RS 232 in the small UAV sector comprise the integration of GPS (global positioning system) receivers, "intelligent" IMUs (inertial measurement units) and compass sensors, wireless data links, and connections between multiple processing units.

Universal Serial Bus (USB)

The *universal serial bus* (USB) interface is, effectively, succeeding the RS 232 port as the standard "desktop" peripheral interface. This is also affecting the design of embedded system components, however, with some due delay. USB has been specified by a group of computer manufacturers, the latest specification document carries the revision number 2.0 [31].

USB uses a 4-wire cable to establish a logical bus, while every physical cable connection is only point-to-point. Thus, any actual configuration has a tree topology, with a single USB *host* as its root, USB *devices* (also called *functions*) as leaf nodes, and – if required – one or more USB *hubs* as interior nodes. Every edge in this tree corresponds to a single point-to-point USB cable connection. In a typical desktop scenario, the host would be the PC, and the devices would be items like keyboard, mouse, scanner, printer, digital camera, or others. USB hubs would provide the necessary branching. Table 3.1 lists some of the features and capabilities established by the most current USB specification.

The description above points out that USB is clearly aimed at desktop applications. With regard to embedded system applications, many of USB's features are useless, and the high complexity of the USB protocol[2] further discourages its use in the latter case. However, it can

---

[2]The standard comprises more than 600 pages. Many of its provisions refer to the protocol and have to be taken into account when designing a USB-compliant device, while the great majority of regulations within the

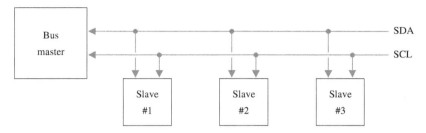

Figure 3.4: Two-wire interfacing scheme underlying the I²C serial bus.

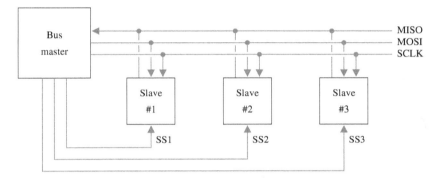

Figure 3.5: Three-wire-plus-chip-select interfacing scheme underlying the SPI serial bus.

be observed that USB interface implementations in sensors and processing units for embedded systems come forth with – and because of – the possible necessity of connecting them to desktop PC systems. This, one the other hand, provides a major issue related to the possibility of employing USB on-board a UAV: Since modern microcontrollers are considered more likely to be used for building USB *devices* than to act as the central processor of an embedded system, they frequently lack the ability to function as a USB host, which in turn would be strictly required of at least one processing unit on board. Consequently, USB must yet be primarily seen as a theoretical possibility more than as a practical option in the area of UAV design. If at all, GPS receivers are the first candidates for USB interfacing, because of their relevance in desktop scenarios.

Serial "Inter-IC" Busses

Serial "inter-circuit" busses are generally designed for the connection of different integrated circuits or circuitry boards within a single device. As such, they correspond more closely to the employment on-board a small UAV than USB. For typical "desktop" issues like hot-swapping, power supply, and very high data rates are not relevant to these busses. There are more than

---

RS 232 standards are *not* truly relevant when implementing a RS 232 interface, for comparison.

a single standardized bus available, but they are dealt with in a common section here because their main characteristics are sufficiently similar.

The main idea behind these busses is the provision of a relatively standardized way of connecting several intra-device components while using a minimum number of chip pins. This is due to the fact that the number of pins tends to be the most relevant cost factor when designing ICs for use as embedded controllers.

One particularly versatile and flexible bus is the I^2C (*inter integrated circuit*) bus designed by Philips in the 1980s [123]. It uses just two lines, called SDA (*serial data*) and SCL (*serial clock*), which are read-write-accessible by all devices connected. Therefore, I^2C only permits half-duplex transmission, i.e. only a single device transmitting at a given time. I^2C can address up to 127 devices connected to the bus. Usually, there is only one device acting as a *bus master*, which means initiating data transfers. But I^2C also supports multi-bus-master configurations with a little extra complexity. Figure 3.4 depicts the interfacing scheme underlying I^2C in the most usual single-slave setup.

Typical data rates for I^2C as defined by the most current specification are 100 kbit/s, 400 kbit/s, and 3.4 Mbit/s. Of these, most available devices support up to 400 kbit/s.

Another widespread bus type is the SPI (*serial peripheral interface*) bus, defined by Motorola. There is no self-contained specification of SPI, instead the bus is defined as a feature of the ICs supporting it, e.g. the M68HC11 microcontroller [97]. SPI represents a compromise with less logical complexity than I^2C, but requiring more lines, their number even depending on the number of connected devices. There are three lines common to all devices, which are

- SCLK (*serial clock*), a clock line always driven by the bus master initiating the transfer,

- MOSI (*master out – slave in*), the output line of the master and shared input line of all slaves, and

- MISO (*master in – slave out*), the input line of the master and shared output line of all slaves.

Therefore, full-duplex transmissions are basically possible with SPI. In contrast to I^2C, there is no way of device addressing defined in the bus protocol. Consequently, there is a dedicated chip select signal (SS – *slave select*) to be provide for each slave device. These signals are used by the bus master to select and communicate to one slave at a time. Figure 3.5 depicts the interfacing scheme underlying SPI.

Possible data rates for SPI are not limited by the specification texts, but rates above 1 Mbit/s are frequently supported by SPI-capable devices. Due to the simple "addressing" scheme and the unidirectional signal transmission, many processing units with some kind of synchronous (i.e. separately clocked) serial interface can work as SPI bus masters by means of suitable software.

Especially SPI is getting more and more common with sensors that may be required in UAVs. There are also ICs available that provide RS 232 interfaces via the SPI bus, e.g. [95]. This permits the design of UAVs with cheap and physically very small microcontrollers as onboard processing units. Acceleration, rotation rate, and magnetic field sensors are the most probable candidate devices for direct SPI interfacing.

CAN Bus

Another popular inter-device bus, designed for longer distances between the communicating devices, is the CAN (*controller area network*) bus. It originates from the European car industry, originally being a response to the vastly increasing number of information processing devices built into modern cars and the pressing demand of minimizing the complexity of intra-car cabling. Meanwhile, it also has gained a substantial market share as an industrial field bus. The CAN bus has been specified through a combination of international standards by ISO [7, 8, 9] and an industrial standard [58].

CAN uses two-wire differential signalling at the physical layer, which permits combinations of data rate and cable length ranging from 1000 m at 40 kbit/s to 40 m at 1 Mbit/s. Single messages are limited to 8 bytes effective length. Arbitration on this multiple-access bus is performed using an arbitration field of up to 11 bit or 29 bit (depending on the protocol version) in each message, which effectively results in a prioritization scheme that guarantees instant bus access for – and only for – messages using the highest arbitration level. Real-time bounds for all remaining levels of priority have to be taken care of in the application design.

While the possibility of very long bus cables is not particularly relevant to small UAV design, CAN bus interfacing is still attractive because CAN controllers implementing many features of the bus in hardware are becoming common in modern microcontrollers. The same holds for "intelligent" sensors. There are GPS receivers with CAN interface available, as well as CAN-to-RS-232 converters. However, devices tend to be predominant still that can usefully be built into cars.

IEEE 802.3 (Ethernet)

Ethernet, basically according to IEEE 802.3 [108], constitutes the most common base for local Internet connectivity and is also gaining market share in embedded devices today. The most obvious reason for that is the fact that embedded web servers for remote device access are becoming more and more popular, and chip manufacturers have started to provide ICs supporting the design of such small web-enabled devices.

For small UAV applications, it is improbable to have suitable position or attitude sensors with an Ethernet interface. But using an Ethernet-enabled processing unit on-board a UAV might be an option for the following two reasons:

- When using wireless Ethernet (IEEE 802.11 [76]) for data transmission to and from a UAV, "wired" Ethernet is one possible way of connecting the processing unit to the radio transmission – via a wireless access-point or similar device on board.

- Certain web-cams are a cheap and often also light-weight option for image acquisition. Especially cameras with built-in pan-tilt and zoom capabilities are frequently equipped with an Ethernet interface in order to act as a stand-alone web server.

It has to be noted that – opposed to CAN, for example – there is no real-time behavior addressed in the common Ethernet standards. Instead, real-time requirements would have to be taken care of by means of a suitable additional software layer, or by the design of the "application" software modules.

3.1.3   Development Support

This section presents features of on-board processing units and the software tools accompanying them that provide valuable support during the development cycle. These features may constitute highly relevant criteria when selecting a processing unit, although they are not required at all in the utilization phase of the designed system.

Development and Build Environment

The development and build environment comprises, at least, the compiler and linker for some target processor. The availability of compilers also determines the set of programming languages that may be used on the processing platform in question. In addition, some development environment may also contain a source editor and tools for project maintenance and module dependency management.

The characteristics, quality, and cost of the development environment have to be thoroughly considered. Especially for prototype design with a very low number of units built, the cost for many development environments may exceed the hardware cost by far. In these cases, open-source development environments, for example the GNU tool-chain available for many target platforms [48, 49], provide a viable alternative to commercial integrated development environments, although it tends to result in less convenient handling from the perspective of modern user interface concepts.

Due to the many different configurations of processing units and circuitry board designs employing them, as well as to the quick emergence of new processor designs and the fact that most development is carried out with a strictly industrial background, there is a high risk of being impeded by inappropriate, faulty compilers – much higher, though, than in desktop software development. Furthermore, the limited memory resources and nonlinear memory models met in the case of microcontroller platforms often require the selection of a certain *memory model* by the software designer, which must be appropriately supported, at least, by the compiler and linker. The availability and support of a certain memory model might easily prove crucial to the suitability of a certain processor type for a certain application. Consequently, it is advisable to perform the basic on-board system design and the selection of one or more processing units in parallel.

One very important magnitude when designing software for embedded systems is the *turnaround time* required after any single change in the software source code. This comprises the time for compiling, linking, possibly building a new bootable system image, and uploading the new software version to the embedded processing unit. Uploading, here, may either mean a single, volatile upload into RAM for a single test run, or programming the new software into some permanent storage medium like EEPROM ("flash") memory. In some architectures, the software image may have to be stored externally on some storage media of the

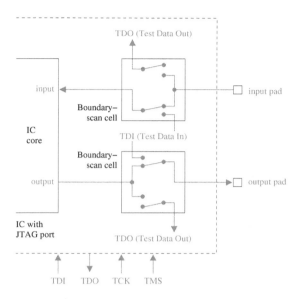

Figure 3.6: Boundary-scan concept implemented via the JTAG port.

second kind, for example a Compact Flash memory card or similar device, the latter having to be subsequently inserted into the embedded system. This great variety of procedures resulting from a minimal software change in the development phase causes an equally great variance of the corresponding turnaround time, which in turn may serve as a relevant criterion in the selection of an embedded processing platform.

Run-Time Monitoring and Debugging

A second issue of embedded system software development, after building and uploading the software, consists in available methods and tools for debugging and monitoring the behavior of the software at runtime. While debugging is conveniently supported in modern development environments for the desktop, this task is much more difficult in the case of embedded systems, primarily due to their lack of suitable "redundant" peripherals. Furthermore, embedded systems operating under real-time constraints cannot usually by debugged by stepping through the software execution because external events the system has to respond to cannot be slowed down accordingly in most cases. This leaves the following catalogue of viable runtime debugging methods in embedded system development:

- *Cross debugging*, using a monitor program on the embedded processor and a debugger running on a desktop machine, connected to the embedded processor via RS 232, for example. This feature is often provided by modern integrated development environments and permits convenient debugging, very similar to desktop software development. It should be noted, however, that real-time characteristics of external events remain problematic and that this approach requires a runtime monitor that fits the actual target board.

- *Outputs* dedicated to debugging purposes. While this may sound non-methodical and old-fashioned, this way of debugging without any tool support is much more reasonable in embedded system development than in other cases. The reason for this is that a stream of debugging output need not interfere with external event timing at all and may permit undisturbed system operation without any prior decision about the next suitable "break point" for state inspection. The only issue to be considered is the need for an unused communications port, such as an RS 232 interface, that can be connected to a desktop PC or some other kind of data display equipment (like, e.g., a traditional serial terminal). Then, it has to be assured that the generation of the debugging output does not interfere undesirably with the execution timing.

- In low level embedded software design, even debugging output is often impossible before proper initialization of the target processor. This renders it impossible to debug the initialization procedure itself. In special cases like this, *line-level* debugging through LEDs connected to some output line of the processor, or even through the measurement of voltage levels via an oscilloscope, may prove surprisingly useful. Even binary decisions about aspects of malfunctions of the software are, naturally, far better than no decisions at all.

- Initiated by the *Joint Test Action Group* (JTAG) and standardized through [107], a simple port has been defined for debugging support in embedded systems. Today, this port is commonly referred to as *JTAG port*. The JTAG port implements *boundary scan*, which means that shift registers separate the signal input and output lines of the IC core from the external pads or pins of the IC – see figure 3.6. There are *boundary-scan cells* located between every external connector and the corresponding line of the IC core. Through some extra logic implementing a simple state-machine-based protocol, the boundary-scan cells can apply some externally supplied logic levels to the IC core inputs or IC socket outputs as well as read out the actual level present there. In this way, JTAG replaces any number of testing pads on the circuitry board and is additionally able to "disconnect" the original connections. For this purpose, a minimum of only four lines at the JTAG port suffices: TDI (*test data in*), TDO (*test data out*), TCK (*test clock*), and TMS (*test mode select*). There are extensions to the standard JTAG capability placing virtual boundary-scan cells inside the IC core, permitting access to internal registers or memory as well. This allows actual runtime debugging via JTAG or the programming of on-chip flash memory. Of course, dedicated debugging devices are required to make use of the JTAG port.

Off-Line Validation

A very reasonable alternative to debugging on the embedded target processor, due to the above-mentioned inconveniences, consists in off-line validation methods. Off-line validation means that the software to be executed on the embedded processor may also be tested externally, i.e. on a desktop computer with far better accessibility from the software developer's perspective. A standard key-word designating such approaches is *software-in-the-loop* testing, meaning that the embedded system's software is executed interacting with the surrounding system components, but not on the target processing platform.

There are two basic options for this kind of off-line validation:

- Running the software inside an *emulator* – this poses the least requirements on the software to be tested, but it does require the availability of a suitable target emulator. Furthermore, the interfacing to external system components may be difficult or even impossible.

- *Compiling* the software natively for the desktop computer. This does not require any specials tools and offers more freedom of interaction. However, special care may have to be taken during software development to support both the target processor's and the desktop processor's programming environment.

Especially the last requirement stated above may be demanding and provides some amount of risk. The latter consists in the possibility that the software works well only on the testing platform and fails when confidently transferred to the actual target system. This issue, of course, strongly depends on the programming language used. Very careful study of the implementation language's specification, its possible pitfalls, and the specifics of all compilers and execution environments involved are strictly required. Using C as a very common programming language for embedded systems as an example, at least the following aspects would have to be taken into account:

- C does not specify the actual storage sizes of the different data types. Instead, the storage sizes depend on the target system and are supposed to match the target machine's command architecture in a certain way. As integer arithmetics in C are never performed with an internal precision of less than the storage size of the type int, implicit type conversions can easily lead to different results on different platforms.

- C lets the compiler implementor decide whether negative signed integer values are right-shifted arithmetically or logically on behalf of the operator >>. This operator is sometimes useful for efficiently dividing by powers of two, which unfortunately requires the shift to be performed arithmetically in order to preserve the dividend's sign.

- The memory layout of variables, especially of struct type, strongly depends on the machine architecture and storage sizes. Frequently, there are very counterintuitive padding rules in place that determine the alignment of individual struct components to memory addresses that are multiples of 2, 4, or even 8. Even for single-valued variables of the same size, one has to distinguish *big endian* or *little endian* byte order, i.e. the arrangement in memory of the least significant to most significant bytes representing the variable's value.

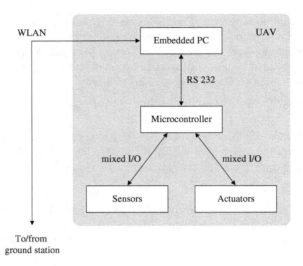

Figure 3.7: Example of an on-board system with *multi-CPU architecture*.

Finally, it has to be noted that cross-platform validation is not limited to classical software programs. MATLAB with Simulink [132], for example, is a system that permits the design of signal processing modules mostly by graphical means. These modules may be executed on an embedded target processor via special *Embedded Target* tools provided by MathWorks, the vendor of MATLAB. The prior experimenting with the signal processing modules on a desktop computer, which is of course common practice for MATLAB users, then fulfills all criteria on off-line validation as discussed in this section.

3.2   CPU

This section deals with the decision on the *number* of processing units to be used on-board a small UAV and with the selection of the actual processor product or products.

The first two subsections, 3.2.1 and 3.2.2, present the basic alternatives of using a multi-CPU architecture or a single-CPU architecture, respectively. Section 3.2.3 then provides an – incomplete – selection of current CPU products possibly useful in small UAV design.

3.2.1   Multi-CPU Architecture

When selecting processing units for a UAV, the following two basic sets of requirements frequently tend to contradict each other:

- The interfacing of sensor and actuator components, as explained above in section 3.1.2, requires processing units with a great number of dedicated input/output features.

- The computational requirements of complex methods of sensor data processing often pose high demands on the computational power on board. Additionally, the algorithmic complexity of these methods seems to recommend the use of processing units "similar" to desktop PCs.

A straightforward solution to this contradiction of requirements consists in the use of multiple processing units: one or several more desktop-like processors for computationally expensive tasks, and one or several microcontroller-type processors for sensor and actuator interfacing. Figure 3.7 depicts a possible architecture of an on-board system utilizing this multi-CPU approach. Here, all high-level data processing is performed by a PC-type processor, while all peripheral hardware is accessed via a separate microcontroller providing sufficient peripheral interfacing capabilities. The coupling of the PC and microcontroller is effected via a standard RS 232 interface usually found in both kinds of processors. Communication to the ground station is performed via wireless LAN directly by the embedded PC in this example; configurations are imaginable where even this communication has to be routed through the microcontroller, too.

The main advantage of the multi-CPU approach is that each subtask is left to a processing unit that is best suited for it. The development of the higher-level software can be conveniently performed solely on a desktop PC. As there are almost no requirements specifically related to the interfaces provided by the PC-type processor on board, the developer is very free in choosing hardware and a suitable operating system. Meanwhile, the software running on the microcontroller is expected to be simple and to undergo only a very limited number of development cycles, due to its trivial input-output structure.

However, it needs to be kept in mind that a system consisting of multiple processing platforms is *always* more complex than one with a single processing unit only. Software *must* be developed for at least two different platforms, requiring two sets of development tools, and changes to the system will often affect both platforms. Besides, every data-stream-type interface between the processors introduces a significant delay into the response time between sensor input and actuator output. When considering the closed-loop system, the worst-case communication delays of the paths both *from* and *to* the microcontroller must be added together. Finally, the communication path also limits the maximum *rate* of sensor input and actuator output. This may turn out particularly critical when continuous input signals overlaid with high-frequent noise are measured by the sensors and the measured signal is supposed to be filtered in software: As a result of the sampling theorem by Shannon and Nyquist, the sampling and transmission rates up to the filter must exceed the doubled frequency of the noise in order to rule out sampling artifacts. Thus, low-pass filtering in software will, whenever necessary, usually have to be carried out by the microcontroller in a multi-CPU architecture. Thus, adjusting the filter may require a change of the microcontroller software.

3.2.2 Single-CPU Architecture

Alternatively, on-board systems may be designed around a single processing unit. The resulting architecture is more flat and less complex. Usually, this requires the use of a

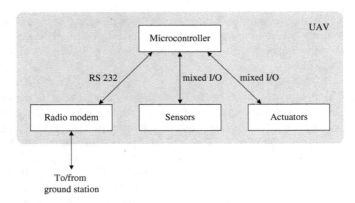

Figure 3.8: Example of an on-board system with *single-CPU architecture*.

	Multiple CPUs	Single CPU
+	Free choice of "high-level" HW/SW	Only 1 software branch
	Desktop-style development	Better control of time constraints
-	Communications overhead	Computational power of μC's
	Added system complexity	More complex μC-style SW development

Table 3.2: Advantages and disadvantages of single and multiple CPU approaches in the design of an on-board data processing architecture.

microcontroller-type processor providing suitable interfacing capabilities. Figure 3.8 depicts a possible architecture of an on-board system utilizing this single-CPU approach. Here, the embedded PC is missing, and the wireless LAN communication is replaced by a radio modem device connected to the microcontroller via RS 232.

The main advantage of this approach is that only a single software branch needs to be developed and maintained. Any time and data-rate constraints can be addressed within a single program, without suffering from the limitations imposed by an additional communications path. However, all requirements, both computational and interface-related, need to be met by the choice of a processor. This makes this choice much more difficult and considerably limits the variety of possible hardware platforms and operating systems.

Table 3.2 summarizes the arguments for and against multi-CPU and single-CPU architectures, respectively. One can state as a conclusion that the main advantages of the multi-CPU approach pertains to the design phase, while the possible real-time performance in the employment stage is superior in the single-CPU case. Thus, multi-CPU on-board systems are mainly justified – and actually prevailing – for prototype design, while single-CPU architectures might be more advisable in a hypothetical series production scenario.

3.2.3 Current Products

This section provides an overview of current processor types and families suitable for use in small UAV on-board system design by means of several examples. For every family of processors, one member product is selected below representing the "typical" performance to be expected from processors of the respective family.

Infineon C166 Family

The C166 family of microcontrollers manufactured by Infineon [12] has found widespread use in the industry, including factory automatization and vehicle electronics. This family, sharing a 16-bit architecture and a common basic command set [11], has now been in the market for some 15 years.

The C166 family is characterized by providing very numerous and flexible ways of peripheral interfacing, combined with sufficient computing power for a variety of applications. Currently, chips of the C166 family are actually designated C161xx, C164xx, C165xx, and C167xx. The common features of most of these products include:

- 25 MHz CPU clock with 2 CPU-clock cycles required for most instructions. Important exceptions are multiplication and division, which are performed 5–10 times slower. Only integer arithmetics are supported.

- Very short interrupt latency due to register bank switching capabilities and interrupt-ability of multiplication and division instructions.

- A dedicated hardware unit named PEC (*peripheral event controller*) is able to handle external events via a single data transfer operation that can be performed automatically without code execution, thus not requiring any context change.

- The controllers can boot via the built-in RS 232 interface, which is useful during development and for re-programming the internal flash ROM.

The SAK-C167CS-4RM, chosen here as the example "top-line" product from the C166 series, exhibits the following list of features:

- 11 KByte internal RAM,

- 32 KByte internal flash ROM,

- 9 16-bit timer-counters,

- 24 10-bit analog-to-digital converter inputs,

- 32 compare-capture-units for signal period measurement and generation, plus 4 dedicated PWM pins,

- up to 111 input/output lines,

- 1 asynchronous serial port (RS 232), 1 synchronous serial port (e.g. SPI), 2 CAN bus ports.

ARM7 Family

The company ARM [92], originally *Advanced RISC Machines Limited*, maintains several processor architectures and licenses them to semiconductor manufacturers for implementing processor products or integrating ARM processor cores into more complex integrated circuits.

The ARM7 family consists of several 32-bit RISC (*reduced instruction set*) processor cores for a wide variety of embedded applications. Offering 32-bit integer operations at up to 130 million instructions per second, ARM7-based microcontrollers provide substantial computing power, while the required die area and the controller's power consumption are very low due to the RISC architecture. ARM7 was introduced back in 1993. Today, resource consumption figures are as good as only 0.26 mm^2 of required die area and 8 mW of power consumption at 133 MHz CPU clock, pertaining to the ARM7TDMI core manufactured in a 0.13 μm process. The developer can choose between a generic 32-bit instruction set for programming or a 16-bit instruction set for increased code density, called *Thumb* [122].

As ARM do not design nor produce microcontroller products themselves, a vendor implementation needs to be selected to present a full actual feature list. This is done here for the Philips LPC2106 microcontroller [106]:

- 64 KByte internal RAM,

- 128 KByte internal flash ROM,

- 2 32-bit timers,

- up to 7 PWM outputs,

- 2 RS-232 ports, I^{2}C bus, SPI port,

- JTAG boundary scan interface.

As the LPC2106 does not provide any external memory bus, it only occupies a very small 48-pin package (7 × 7 mm^2). The company MCT, Berlin, offers a tiny OEM module carrying the LPC2106, which fits into a standard 32-pin DIL socket (18 × 40 mm^2).

The ARM7 core is supported as a compilation target for the free C compiler GNU gcc [48].

ARM9 Family

ARM9 is a more advanced processor core family than ARM7. The main advantage is that there are coprocessor cores available for floating-point arithmetics (FPUs). Besides, ARM9 cores provide more computing power than ARM7, also without an FPU, of up to 300 million instructions per second. Furthermore, some cores of the ARM9 family provide an extended instruction set supporting DSP (*digital signal processor*) features [122].

Freescale/Motorola MC683xx

In in the wake of the well-known MC68000 processor by Motorola, the MC683xx family transports its core architecture and command set into the realm of microcontrollers. Due to the 32-bit core design, computing power is easily sufficient for the majority of UAV on-board system tasks. Today, the vendor of these processors has changed the company's brand name to *Freescale Semiconductor*.

There is an interesting design decision associated with the 683xx series: Most of the peripheral interfacing capabilities are integrated into a very flexible functional unit called TPU (*time processing unit*). The TPU consists of 16 mostly independent channels, each of which can be assigned one of a great variety of complex functions and owns an individual input/output pin. These functions include an RS 232 transmitter or receiver, different kinds of PWM input or output (one complex example being *period measurement with missing transition detection*), stepper motor control, and others. These functions are implemented through microcode and may be extensively configured. There are different versions of the TPU available in different microcontroller products, but the basic approach is common among them.

One somewhat older but still highly popular product from this series is the MC68332 micro-controller [50]. It exhibits the following features:

- 68k core with up to 25 MHz CPU clock,

- 2 KByte internal static RAM,

- up to 24 digital input/output pins,

- TPU with 16 channels/pins, including 16 timer-counters,

- RS 232 port through QSM (*queued serial module*), TPU may provide up to 16 RS-232 lines (transmitters *or* receivers) in addition,

- SPI port.

Another, more recent product of this line is the MC68F375 microcontroller [51]. It provides additional peripheral connectivity and flash ROM, the latter permitting single-chip applications with no external data bus required. Its features include:

- 68k core with up to 33 MHz CPU clock,

- 14 KByte internal RAM,

- 256 KByte internal flash ROM,

- 8 KByte internal late-programmable ROM,

- up to 48 digital input/output pins,

- TPU with 16 channels/pins, including 16 timer-counters,

- analog-to-digital converter with 16 channels, 10 bit resolution,

- 2 RS-232 ports, SPI port (plus TPU functions).

Freescale/Motorola 68HC11 Family

The family of 68HC11 8-bit microcontrollers has found very widespread use in simple embedded systems and, especially, as peripheral interfacing controllers in multi-CPU setups (see section 3.2.1). In this kind of application, 68HC11 controllers may even be used in single-chip mode without external memory, despite their very limited internal memory resources.

A high-end representative of this family is the M68HC11E1 controller with the following list of features:

- 68HC11 core, variable clock speed 0–3 MHz,

- 512 Byte internal RAM,

- 512 Byte internal flash ROM,

- analog-to-digital converter with 8 channels, 8 bit resolution,

- 8 capture and compare lines (simple PWM input or output) with 16-bit timer-counter,

- RS 232 port, SPI port,

- up to 38 digital input/output pins (shared with higher-level peripheral functions).

Atmel AVR 8-bit Family

The family of 8-bit "RISC" microcontrollers by Atmel [29] provides a huge range of powerful 8-bit microcontrollers with integrated system memory, all of them basically code-compatible, with pin-outs ranging from only 8 to 100 pins. The instruction set provides 8-bit single-cycle arithmetics and has as much as 32 general purpose registers at its disposal.

The Atmega168 [30] shall serve as an example of this attractive family. Being available in a 28-pin DIP version, it is very well suited for prototype design, and its performance and interfacing capabilities render it perfectly suitable for a range of single-ship control tasks all by itself:

- clock speed up to 20 MHz (8 MHz without external oscillator),

- 16 KByte flash program memory (self-programmable),

- 1 KByte Byte SRAM,

- 512 Byte EEPROM (mainly for persistant data storage),

- up to 23 digital input/output pins,

- 8 10-bit analog-to-digital converter inputs,

- RS 232 port, SPI port.

3.3 Operating System

This section discusses the available options in choosing an operating system, or even more generally, in providing the tasks concerning system operation as identified above in section 3.1.1.

While it is quite obvious that an operating system is strictly required in the desktop sector in order to provide the necessary abstraction for running application software, things are not quite so with embedded real-time systems, or small UAV on-board systems as a special case. Just to point out some of the relevant differences of embedded systems:

- There is no direct user interaction, at least no classical desktop-style user interface, in the case of an embedded system.

- It is often possible to perform static scheduling of the tasks to be executed, for the data flow and the set of tasks are well known a priori. This is particularly valid for basic control tasks.

- Due to the static nature of the tasks, dynamic resource management may also be dispensable.

- Services like file system access and standardized networking services are often unnecessary as well.

On the other hand, real-time bounds for the reaction to sensor inputs, commands sent from the ground station, and other external events need to be strictly guaranteed, due to the substantial amount of possible damage associated with system malfunction in an airborne vehicle (see [91] for in-depth coverage). None of the common desktop operating systems does offers any kind of real-time guarantee. Thus, with regard to operating systems there are two possible choices:

1. to use an off-the-shelf, usually commercial, operating system providing real-time guarantees, or

2. Not to use any off-the-shelf operating system at all.

Here, the second alternative may sound surprising to system developers accustomed to modern, complex, high-performance computing devices. However, as a part of this section it will be explained why, how, and on what kind of processing units it may actually be advantageous to provide all "operating system level" tasks as a part of the particular "application" software, without using any dedicated operating system at all.

The remainder of this section will first provide a short introduction to some of the off-the-shelf operating systems suitable for small UAV on-board computers (section 3.3.1), then address the approach of using no separate operating system at all (section 3.3.2).

3.3.1   Current Products

There a great number of real-time operating systems available commercially, and this section can only discuss a very limited – and somewhat arbitrary – selection of them in greater detail. The following list mentions several products, but only a few of them are treated in separate sections below:

LynxOS by Lynx Real-Time Systems, Inc. [94]: POSIX 1/1b/1c compliant and similar to UNIX, available for x86, Motorola/Freescale 68k, PowerPC, Sun SPARC, and other architectures.

Microware OS-9 by RadiSys Corporation [35]: "UNIX-style" real-time kernel, available for x86, ARM, PowerPC, Motorola/Freescale 68k, MIPS, and other architectures.

QNX Neutrino RTOS by QNX Software Systems [128]: POSIX compliant microkernel operating system, available for x86, ARM, MIPS, PowerPC, Hitachi SuperH, and other architectures. Covered in more detail in section 3.3.1.

RTLinuxPro by FSMLabs, Inc. [53]: Provides an extra real-time layer beneath a standard Linux system, available for x86, PowerPC, ARM, MIPS, and other architectures. Covered in more detail in section 3.3.1.

VxWorks by Wind River [117]: POSIX-compliant microkernel, available for x86, ARM, PowerPC, MIPS, and other architectures.

Windows CE by Microsoft Corp. [32]: Operating system that Microsoft's *Mobile* product range is based on. Available for x86, ARM, PowerPC, MIPS, Hitachi SuperH, and other architectures. Covered in more detail in section 3.3.1.

QNX

QNX is a full UNIX-like operating system with real-time capabilities. The vendor even provides a fully integrated development environment running under QNX on a desktop PC (QNX Momentics), which can prove advantageous also for certain software-in-the-loop testing scenarios.

As QNX natively supports the POSIX standard [77] including many non-obligatory options, the porting of any available source code written for some version of UNIX is usually effortless and simple. In many cases, such code will run "out of the box" under QNX. The author has joined in porting parts of the MARVIN UAV [68] communications software to QNX on one occasion[3]. In this case, the device file names used for accessing the RS-232 ports were basically the only aspect that had to be adopted from a "desktop" POSIX implementation.

A specialty of QNX is that every peripheral port is handled by a separate system process running in user space. Therefore, the QNX image has to be configured to start one such process for each of the RS-232 ports to be used, for example.

---

[3]Not published due to a non-disclosure agreement.

A more elaborate list of the features of QNX includes:

- microkernel with driver implementation in user space, based on message passing.

- Extensive POSIX 1003.1-2001 [77] support including real-time extensions and threads.

- Transparent support for symmetric multiprocessing (SMP).

- Graphical user interface QNX Photon, for target *or* development system.

- *FIFO*, *round-robin*, and *sporadic* scheduling policies.

- Thread priorities ranging from 0 to 63 (highest), with *priority inheritance* protocol implemented to prevent *priority inversion* [91]. Otherwise, priorities are static for the FIFO and round-robin scheduling options.

- All POSIX thread synchronization services supported: mutexes, condvars, barriers, sleepon locks, reader/writer locks, semaphores, message send/receive/reply, atomic operations.

All in all, QNX constitutes a convenient and powerful real-time operating system, especially when software porting from, or cross-platform interfacing with, other POSIX-based operating systems needs to be addressed.

RTLinuxPro

Real-Time Linux, or *RT-Linux*, follows an approach completely different from "native" real-time operating systems like e.g. QNX: With RT-Linux, a standard Linux kernel, together with all user processes, runs as the lowest-priority thread of a small and dedicated real-time kernel. For the actual real-time tasks of the application, other threads are created running under the control of the real-time kernel. In the case of RT-Linux, these threads are implemented as Linux kernel modules, as is the whole real-time kernel. The decision of using kernel modules for this purpose is straightforward, because it constitutes the standard way of adding kernel code to a Linux system at runtime. When the real-time kernel is executing, the standard kernel runs embedded in a virtual machine layer so that the interrupts originating from this virtual machine cannot affect real-time performance at all [19].

The big advantage of this approach is that high-level applications and user interface libraries well known to the Linux developers and users can be used without any change. Thus, applications like a KDE desktop or the Mozilla web browser may be used on a machine engaged in real-time control. Tasks with real-time bounds assigned to them are not affected by these high-level processes. In addition, the real-time kernel can be very small, because aspects like user interface and general file-system services do not need to be addressed there.

The downside of this approach, however, is that there is a rather limited programming interface (API, *application programming interface*) provided to the real-time tasks because the latter cannot be permitted to access the standard Linux kernel's services. Second, the developer is required to observe that all interfacing services provided by the standard kernel must be completely kept out of real-time considerations and guarantees. In practice, this will often require to limit any guarantees to immediately safety-critical aspects only – aspects that do neither involve user interaction nor remote parts of a distributed system.

The original version of RT-Linux stems from academic background and is available under open-source conditions [5, 19]. Meanwhile, the company FSMLabs, Inc. [53] has performed a major redesign of RT-Linux, resulting in RTLinuxPro, which is a commercial product. FSM-Labs also own "RTLinux" as a registered trademark and offer access to the version RTLinuxFree under free-software licensing (GPL).

While the original ("free") version offered a small and proprietary API for the real-time tasks, the commercial version has adopted a subset of the POSIX standard [77] for alternative use. This makes the application of RT-Linux much easier for developers with prior experience in POSIX real-time programming. Additionally, features like limited file-system semantics, user space memory access from real-time tasks, and real-time task implementation in `main()` function style (instead of as a kernel module) have been added.

The code example shown by figure 3.9 shall depict the style of RT-Linux real-time programming. It provides a small task that reads data from an analog-to-digital converter at a fixed rate of 2000 Hz and passes these data to some user process by means of a so-called *real-time FIFO* channel. This example is basically taken from [143], with the error handling sections removed to emphasize the basic approach only. In this example, the device file name `"/dev/rtf0"` selects the desired FIFO, and the `pthread_*` calls are POSIX-compliant.

For the x86 architecture, FSMLabs state worst-case[4] thread latencies ranging from 13 μs for an AMD Athlon SMP system (at only 1.2 GHz CPU clock) to 104 μs for the *Prometheus* PC/104 by Diamond Systems.

Windows CE

While primarily an operating system for hand-held computers and mobile devices, *Windows CE* [32] does offer response time guarantees and all services required for embedded real-time system implementation.

The current version at the time of writing is Windows CE 5.0. There are some product names closely related, but not corresponding, to Windows CE: *Windows Mobile* refers to a set of software packages for the operation of certain classes of hand-held ("mobile") devices that commonly run Windows CE as an operating system, and *Pocket PC* is a definition by Microsoft of such a class of hand-held devices running a *Windows Mobile* software suite.

The features of Windows CE include:

1. *Round-robin* and *FIFO* scheduling policies possible, the latter one by setting the time-slice (*thread quantum*) to zero. By default, *thread quantum* amounts to 0.1 s.

2. Thread priorities ranging from 0 to 255 (lowest), with *priority inheritance* protocol implemented to prevent *priority inversion*. Otherwise, thread priorities are static.

---

[4]Determined experimentally, however.

```
// global declarations
pthread_t T;
int       fd;
int       stop = 0;

#define   DELAY_NS 500000 // 500 microseconds delay

// task function
void *my_code(void *arg)
{
  struct timespec t;
  struct mydata D;

  clock_gettime(CLOCK_RTL_SCHED, &t);
  while(!stop) {
    copy_device_data(&D.d);
    write(fd, &D, sizeof(D));
    timespec_add_ns(&t,DELAY_NS);
    clock_nanosleep(CLOCK_RTL_SCHED, TIMER_ABSTIME, &t, NULL);
  }
  return (void *)&stop;
}

// initialization function
int init_module(void)
{
  fd = open("/dev/rtf0", O_WRONLY | O_NONBLOCK | O_CREAT);
  pthread_create(&T, NULL, my_code, NULL);
  return 0;
}

// termination function
void cleanup_module(void)
{
  stop = 1;
  pthread_join(T, NULL);
  close(fd);
}
```

Figure 3.9: Example code (excerpts) of an RT-Linux task that reads data from an A/D converter and writes them to an RT FIFO. Error handling would also be necessary but has been left out here for clarity reasons.

3. Synchronization services include critical sections, mutexes, events, and semaphores.

4. Minimum *memory footprint* of operating system is about 350 KByte only. The system can be freely configured by the device manufacturer.

5. Abstraction from the actual hardware and CPU platform is provided via the *OEM abstraction layer* (OAL), which has to be built by the device manufacturer.

6. Two-stage interrupt processing: A very short *interrupt service routine* running in system mode triggers the final handling through an *interrupt service thread* (IST), which is basically running in user space.

3.3.2   No Universal OS

After discussing some possible choices of an operating system for an on-board processing unit in the previous section, this section suggests still another alternative: not to use any off-the-shelf operating system at all, but to include just the minimally required "operating system" services inside the application software.

This kind of approach is clearly not a possible choice with desktop computer systems – these *must* rely on common programming interfaces and the usability of common software products. However, in the case of embedded real-time systems in general and UAV on-board computers in particular, there are a fixed structure of the tasks to be performed (see figure 3.1) and a fixed structure of the communication channels to be serviced. Furthermore, the hardware configuration of such on-board processing devices is usually much more application-specific than that found in desktop scenarios. These facts make it both easier and more desirable to get by without a separate operating system.

A clear advantage of not using an operating system is that assuring real-time bounds is much easier and does not depend on the reliability of the assertions stated by the operating system's vendor. FSMLabs, for example, the manufacturer of RT-Linux, state worst-case latency times that have been determined through long running *experiments*, as mentioned above in section 3.3.1. This is certainly a good effort, but not formally sufficient. For testing, as is well known, can never prove anything but the possible *violation* of a certain given time bound in this case. The huge combinatorial complexity of a multi-process real-time system yields a vanishing low probability that the actual worst-case timing will actually ever occur during any test.

Furthermore, using an all-purpose operating system always adds redundant complexity to the resulting real-time system, because every application will only use a limited subset of an operating system's services and features, while the executed operating system software code still addresses these actually redundant features as well. In contrast to that, integrating the operating system services into the real-time application enables the system developer to exactly determine and analyze the actual worst-case execution paths of the real-time system.

The main tasks relevant to a UAV on-board system and usually performed by an operating system are the following:

Scheduling, i.e. the allocation of the resource *CPU*, is the most basic function of an operating system. It is very complex as soon as arbitrary kinds of tasks are possible and real-time bounds have to be ascertained. On the other hand, the tasks met in an UAV on-board system performing flight control are very much data-driven, as has already been pointed out in the context of figure 3.1 above: Data has to be read from sensors, possibly transformed or filtered in various ways, fed into the controller, and results in control inputs to be applied to some actuators. Obviously, this scheme of tasks is well suited for *rate monotonic* scheduling [91]: all tasks or "processes" run at some fixed and identical rate of CPU assignments. Thus, the different tasks' code sections can easily be invoked sequentially in a single loop. Whenever the scheduling requirements in an on-board system can be fulfilled that easily, the use of a separate operating system is probably superfluous.

Communications, i.e. the exchange of data between one or several processors on board and one or several processing units on the ground, is the second generic task associated with on-board system design. While the implementation of communication protocols and services is tedious, communication requirements in distributed UAV system involve real-time characteristics, the interconnection of heterogeneous component systems, and possibly the creation of a non-distributed abstract view of the system (refer to 3.1.1). Usually, off-the-shelf operating systems do not provide any communications service that meets all these requirements. Therefore, the benefit from using an operating system tends to be limited to the lowest levels of communication (i.e., OSI [44] layers 1 and 2). For a single communications port, e.g. RS 232 or SPI, the implementation of these layers is not so much costly to justify the use of an operating system just for this reason. In order to meet the more complex requirements mentioned, some *middleware* layer will usually have to be provided – this issue is not covered here, but will instead be discussed in detail in chapter 4.

MARVIN, the UAV system developed with participation of the author, is an example of not using an operating system on board the UAV [68, 102]. MARVIN's on-board computer is an SAB80C167 microcontroller [12] running a single-threaded C program.

Figure 3.10 depicts a simplified version of MARVIN's on-board main-loop code, which alone is responsible of task scheduling. After initialization, the while(1)-loop shown constitutes the outer loop of control flow. In every cycle of this loop, called a *control cycle*, all task procedures *_loop() are called sequentially[5]. Then, the call to bb_sync() triggers the cross-system data exchange necessary for command reception and data logging – this will be covered in more detail in chapter 4. Subsequently, the time elapsed during the current control cycle is calculated in the variable looptimer (in ticks of the main timer T1, the frequency of which in Hz is indicated by the macro T1_FREQUENCY). With CONTROL_FREQUENCY expressing the desired rate of main-loop execution f_C, the quotient assigned to load_wert equals the CPU load in percent of the current control cycle, i.e. the relation of the elapsed time and the maximum permitted time for a single execution of the series of task procedures. Finally, there is

---

[5]The STOPWATCH macro calls are for profiling purposes, they expand to nothing in the case of normal operation.

```
// declarations
unsigned long looptimer;        // time elapsed in this cycle
unsigned long looptimer_old;    // (nominal) cycle start time
int           load_wert;        // current CPU load in %

...

// call task initializations
init_Air ();

// main loop
while (1) {
  // call task functions for this cycle
  led_loop ();                 STOPWATCH(" LED_loop\n" );
  gps_loop ();                 STOPWATCH(" GPS_loop\n" );
  imu_loop ();                 STOPWATCH(" imu_loop\n" );
  compass_loop ();             STOPWATCH(" compass_loop\n" );
  PosFilter_loop ();           STOPWATCH(" PosFilter_loop\n" );
  heli_loop ();                STOPWATCH(" heli_loop\n" );
  servos_loop ();              STOPWATCH(" servos_loop\n" );

  // trigger data exchange with ground station
  bb_sync ();                  STOPWATCH(" bb_sync\n" );

  // determine time consumed in this cycle
  looptimer =  getexpandedT1() - looptimer_old;
  // calculate CPU load in this cycle in %
  load_wert = looptimer / (T1_FREQUENCY/CONTROL_FREQUENCY/100);

  // check time consumed against full period
  if (looptimer <= T1_FREQUENCY/CONTROL_FREQUENCY) {

    // <= 100% CPU load: busy waiting for end of time-slice
    while(looptimer < T1_FREQUENCY/CONTROL_FREQUENCY) {
      looptimer =  getexpandedT1() - looptimer_old;
    }
    // increase nominal cycle start by exactly one period
    looptimer_old += T1_FREQUENCY/CONTROL_FREQUENCY;

  } else {

    // TIME-SLICE OVERFLOW: only reset cycle start to now
    looptimer_old = getexpandedT1();

  }
}
```

Figure 3.10: Static rate-monotonic task scheduling in the MARVIN on-board software.

busy waiting performed until the nominal control period $1/f_C$ has fully elapsed. Here, busy waiting does not provide any problem since there is no concurrent process that could use the CPU time consumed for it. By incrementing `looptimer_old` by exactly the nominal control period in timer ticks, it is assured that the nominal rate of main-loop executions is exactly achieved in the limit (`looptimer_old` contains *approximately* the current time just after being incremented). This only holds, however, as long as the permitted maximum cycle time is never exceeded within the task procedures.

Obviously, each task procedure must return within some time bound limiting the total execution time of one loop cycle to $1/f_C$. Blocking is not possible anywhere in the task procedures because there is no effective concurrency at all. For data-flow-oriented control tasks, this way of processing is perfectly normal. In the case presented here it may even be advantageous, because the choice of a fixed optimal sequence of task invocations (input, processing, output) reliably minimizes the timing jitter between producer and consumer tasks, which might be an issue with multiple processes independently scheduled by some operating system.

f_C has to be an upper bound of all desired task execution rates[6]. Tasks supposed to run at a lower rate than f_C would have to either skip their code in some of the respective task procedure calls, or to run more often than strictly required. Both is easily accomplished and does not add significant programming complexity. With regard to the real-time bounds, it suffices to prove that the total execution time of all tasks when *not* skipping execution does not exceed $1/f_C$. It is comparatively easy to measure the worst case execution time of all task procedures without stochastic testing, by just enforcing worst-case branching in a single run.

In the MARVIN case, $f_c = 40$ Hz. Execution rates of this magnitude with control tasks should be manageable by most modern microcontrollers. Concerning the flight control of small helicopters, it is interesting to note that higher control frequencies are largely useless, since the main rotor can exert the roll and pitch control torques only alternately, alternating at the doubled rotation rate of the rotor. For MARVIN, resulting from 1350 RPM of the main rotor, this alternating frequency amounts to only little more than 40 Hz.

Whenever a "process" has to be executed that is not per se periodic, or the periodic execution time of which excesses the time permissible for it per cycle, its task procedure must be programmed to return after partial execution and to resume its execution on the subsequent call. This amounts to implementing the task procedure in the form of a state machine. This can be done, but it does finally add substantial inconvenience to the development process. In the latest when this applies to more than one process, it is worthwhile considering the use of an operating system to solve the scheduling problem, though.

For peripheral interfacing, the control flow of the task procedures described so far is complemented by interrupt service routines. These routines are very small and communicate with the user-level software through global variables. Only for a very limited number of short copy operations regarding some of these variables, the user-level software needs to disable interrupt processing. This increases the worst-case interrupt latency time by the execution time of the longest section with interrupts disabled, which can be easily determined. Some of the interrupt service routines, for example the one for reading the analog-to-digital converter's output, are

---

[6]Formally, of course, their least common multiple (LCM). But this will often be eligible for relaxation, depending on the maximum permissible delays applicable to the processes in question.

triggered by hardware timers of the controller and exhibit an execution rate much higher than f_C.

3.4   Subsumption

This chapter has examined the available choices of on-board processing units and related architecture issues.

As every on-board system constitutes a network of sensors and processing units, the related requirements on computing and interfacing capabilities are networked as well and demand a global design solution. Changing a single component later might only be possible at the expense of a full system redesign.

Regarding the number of computing nodes as well as the operating software utilized, it is obvious that there is a distinct conflict between desirable abstraction layers on the one hand and unnecessarily high run-time resource requirements on the other. While this kind of decision usually favors design abstraction with desktop-class systems, a highly specialized small-footprint solution like a singe-microcontroller no-operating-system design may actually be favorable with a small UAV system. This results from the specialized operating conditions and several extra constraints regarding size, weight, and power consumption.

Chapter 4

Communications

In the design of any distributed embedded system, the task of communication between different sources and sinks of information has to be solved. This involves both dedicated hardware and dedicated software to provide a suitable communications service. For the corresponding software layer, the term *middleware* (see e.g. [4]) has been coined, indicating that communications occupy an intermediate role between the local low-level "operating system" tasks and the distributed "application" governing the distributed system. Hence, as already indicated in section 3.1.1, the middleware and communications layer provides a new abstract view of the system, hiding the issue of distribution from the developers of the high-level software modules.

With the focus on small UAV systems, it turns out that the set of requirements on the communications service is equally typical of small UAVs as this is the case for requirements existent in other design fields – see chapter 1. Notably, both the real-time system properties and the hardware diversification met in such systems tend to be poorly addressed by commercial middleware products. Of course, attempts are being made to address middleware issues "once and for all", e.g. through the *Common Object Request Broker Architecture* (CORBA) [64] with its real-time extension [63], but these suffer heavily from their out-of-hand complexity, and they still do not specifically address the real-time requirements of low-level data exchange procedures, as will be pointed out in the current chapter.

This chapter starts, as usual, with the identification of communication requirements in small UAV design in the subsequent section 4.1. Then, section 4.2 addresses the options in choosing suitable devices and technologies for low-level data exchange, i.e. "communication hardware". The following two sections discuss two opposing alternatives in the choice of middleware, which are the CORBA standard in section 4.3 and the Blackboard Communications System (BBCS), originally developed for the MARVIN UAV system, in section 4.4.

4.1 Requirements

This section lists the requirements on the communication infrastructure that are typically involved in the design of small UAV systems. These can be separated into requirements on the transmission "hardware", i.e. the technology used for transmitting arbitrary data between system components with no material links between them, and requirements on the communication software that uses above transmission devices to provide a more abstract view to the software developer – i.e. the employed *middleware*.

Transmission devices for use on-board a small UAV should exhibit:

- Naturally, among the main requirements are suitably low *weight* and *power consumption* for use on board.

- Second, *reliability* is particularly critical because any permanent loss of communication will cause an immediate risk of property and bodily damage here.

- As many heterogeneous electronic systems operate in very close proximity on-board a UAV, the issue of *interference* is highly relevant. Particularly, the use of several different radio links, maybe plus a GPS receiver, strictly requires a thorough examination of the system's susceptibility to cross-component interference.

- *Cost*, while usually just one out of many criteria, is somewhat special here because the cost of equipment may be complemented by substantial cost of operation here whenever a public network of any kind is used. Using a public network can be very advantageous, if not obligatory, considering very long-range missions.

- Finally, regarding the "autonomy" of a UAV commanded via radio links, care should also be taken to provide *security* against possible malicious attempts to gain control of the UAV.

The typical set of requirements on the *middleware* software layer includes at least the following list of aspects:

1. There are very different pieces of information to be transmitted between pairs of system components. The difference pertains to the size of information units, their production (or update) rate, and the upper bound on the delay of their being relayed as imposed by the real-time guarantees to be met by the system.

2. The available outgoing communication bandwidth from every component needs to be shared between these different pieces of information transmitted by the component in question. This sharing, in turn, affects the resulting transmission delays.

3. A great fraction of the information to be exchanged consists of *state variables*. State variables shall denote data portions of a fixed size, with only the most recent content of the variable being of relevance. I.e., in the exchange of state variable contents, the loss of data (or failure to schedule a transmission within a certain time window) is fully cured by the next successful transmission. Therefore, the use of flow control or means of retransmission may be counterproductive.

4. The middleware might have to interconnect heterogeneous components based on a great variety of processing hardware and operating systems. If this includes the implementation on microcontrollers, a relatively light-weight middleware system is advantageous.

4.2 Transmission Devices

This section provides an incomplete overview of communication devices and technologies suitable for low-level data exchange to and from small UAVs. The peer endpoint of communication, in this context, may either be the ground station, as certainly present in all UAV systems of practical use, or even another UAV in the case of more complex systems comprising multiple aerial vehicles.

Low-level communication, in this context, shall pertain to any point-to-point or point-to-multipoint transmission of a data stream, not necessarily including flow control and the detection and correction of transmission errors. In others words, the devices and supported protocols discussed here need to provide at least the services covered by layers 1 and 2 (physical and data link layers) of the OSI reference model [44]. Optionally, support of layers 3 and possibly 4 (network and transport layers) may be present, but might overlap with the middleware's services anyway and is thus not required.

Not surprisingly, all technologies to be considered here are based on radio frequency (RF) transmission.

4.2.1 Remote Control Sets

Stemming from the sector of remotely piloted model aircraft, RF remote control sets are a very obvious equipment of small UAVs. Particularly UAVs based on some model aircraft will carry a remote control receiver from the factory, and it is often helpful to keep it for testing and emergency recovery purposes. Being commercial products, remote control receivers can be considered reliable, compared to on-board system components just under development.

Another good reason for keeping the remote control receiver is that the remote control transmitter, with two two-axis joysticks usually, provides a well-proven user interface for steering aerial vehicles. In most cases of small UAVs, it is clearly desirable to offer an assisted remote control mode, with a human pilot easily commanding speed and direction with these joysticks, while on-board flight control takes care of stabilization and ensures safe behavior whenever the pilot fully releases the controls. The implementation of this features naturally requires that the on-board computer is capable of reading the control signals sent via the remote control set and controlling the UAV's flight-relevant actuators at the same time.

In this special case, data transmission is one-directional and does not consist of an arbitrary data stream, but of a vector of a small number (like 4–12) of more or less analog control signals. The usual way of PWM coding of remote control signals has already been set forth in section 3.1.2.

4.2.2 Radio Modems

A pair of radio modems provides wireless data transfer usually with RS-232 or similar interfaces on both devices. There are many different products available, and the kind of signal modulation scheme used may have a substantial impact on the transmission quality and susceptibility to interference. The author has once experienced intolerable interference caused even by the running two-stroke engine of a model helicopter, so selecting a product only based on its data-sheet, without practical testing, must be considered utterly inadvisable.

Due to the great number of products, it is not meaningful to attempt a comparative listing. Therefore, only two example products are arbitrarily chosen here. The first one is the *dataTaker 905U-D High Speed Industrial Radio Modem* [93]. This is a list of some of its features, referring to the version legal for use in Europe:

- RS 232 or RS 485 standard serial interface,

- data rates 19200 bit/s, 57600 bit/s, 76800 bit/s selectable,

- 500 mW transmit power,

- line-of-sight transmission range, maximum 5 km @ 19200 kbit/s,

- power consumption 350 mA @ 12 V.

The second example, targeted at somewhat higher transmission performance, be the *FreeWave FGR-HTPlus Industrial Radio* [52]. This device is not intended for use in Europe. Its features include:

- both Ethernet and RS 232 connectors provided,

- up to 540 kbit/s link throughput,

- spread spectrum technology,

- 100 mW to 1 Watt transmit power,

- line-of-sight, point-to-point transmission range up to 66 km,

- power consumption when transmitting max. 550 mA @ 12 V, max. 140 mA when only receiving,

- mass 427 g including enclosure.

4.2.3   Wireless LAN

Wireless LAN (*local area network*) according to IEEE 802.11 [76] and subordinate standards is the most common technology for short-range high-speed wireless data exchange today. The currently most relevant standard version is IEEE 802.11g [109]. Its main features are:

- supported data rates of 1, 2, 5.5, 6, 9, 11, 12, 18, 24, 36, 48, and 54 Mbit/s with *adaptive rate selection*,

- transmission frequency 2.4000 GHz–2,4835 GHz,

- transmit power limit (in Europe) 100 mW = 20 dBm EIRP.

The resulting outdoor range strongly depends on the antenna type used. As a general rule, up to 300 m may be possible outdoors even with external omnidirectional antennae but at a reduced data rate between 2 and 11 Mbit/s. While high-gain directional antennae may enable line-of-sight ranges up to 10 km, this is not really relevant to UAV applications due to the stringent line-of-sight restriction and the need of continuous alignment of the antennae at both endpoints. As there are only 3 strictly non-overlapping channels available, the use in multi-UAV systems or in the potential presence of additional wireless LAN transmitters outside the system is further aggravated.

It has to be noted that the high available data rate is unique to wireless LAN, so that it is very attractive for high-bandwidth communication tasks such as image and video transmission. Furthermore, the broad availability of devices at very low cost is appealing. However, due to the usage and range limitations mentioned above, it is not advisable to employ wireless LAN for safety-critical messages or any kind of messages with associated time bounds.

4.2.4 DECT

DECT, or *digital enhanced cordless telecommunications*, is a standard basically intended for digital cordless telephones. It has been officially devised by ETSI as EN 300 175-1 [42] and a series of subordinate standards. Meanwhile, the standard has been adopted by a great number of vendors for the design of data communication devices. The wide distribution of DECT telephones conveniently minimizes the cost of the corresponding low-level transmission circuitry.

Features of the DECT standard that are common to telephones and data transmission devices include [47]:

- 10 distinct carrier frequencies (channels) between 1880 Mhz and 1900 MHz,

- time-division multiplexing with 24 uni-directional timeslots per channel,

- 32 kbit/s data rate per timeslot and direction raw, 24 kbit/s secured,

- channel bundling permits up to 552 kbit/s of secured data transfer,

- dynamic channel selection and handover capabilities, optimized for "hostile" radio environments,

- maximum peak transmit power 250 mW, maximum average transmit power 10 mW.

From these specifications, a nominal outdoor transmission range of up to 300 m results, despite the very low transmit power limit. In line-of-sight cases, 800 m are achievable. With respect to the reliability of the transmission, application of DECT to UAV communications seems reasonable up to 300 m of operating range for mission-relevant data.

Figure 4.1: DECT data transmission devices *Gigaset M101 Data* by Siemens (left) and *HW86010* by Höft & Wessel (right).

A number of standardized *data service profiles* (DSP) are under development by ETSI. These standards allow the inter-operation of data transmission devices from different vendors. However, proprietary implementations of data transmission devices have been available in addition, at least since 1999.

Figure 4.1 shows two different devices actually used in UAVs developed with participation of the author. The *Gigaset M101 Data* by Siemens [72] has been the first mass-market component for data transfer using DECT. It offers a transparent point-to-point RS-232 connection at up to 115200 kbit/s and constituted the primary data link in the 2000 "Mark I" MARVIN system [68, 99]. It has proven reliable beyond expectation during several years of practical on-board use, above range figure of 800 m primarily stemming from this device. The *HW86010* by Höft & Wessel [10], on the other hand, constitutes an OEM communications module with external antenna connector and is thus much better suited for system integration. The core module weighs only 30 g. It provides about the same capabilities as the M101 with RS-232 interfacing, but up to 500 kbit/s data rate in principle. The HW86010 has been used for MARVIN "Mark II" [68, 115]. Unfortunately, its reliability in the field regarding drop-outs has turned out to be clearly inferior to the older Siemens device.

4.2.5   GPRS

As all communication options discussed so far exhibit severe range limitations that would impede many desirable applications, a suitable resort consists in using a public wireless network for data exchange. Considering both current technology and current network fees, the data transmission service GPRS (*General Packet Radio Service*) [43] available in all common GSM mobile phone networks is the most viable option.

GPRS constitutes a packet-oriented mobile data transmission service. Charging usually occurs according to the transmitted data volume. The transmission rate is theoretically limited to 171.2 kbit/s for pure uni-directional transmission, which would require the concurrent use

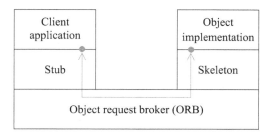

Figure 4.2: CORBA core structure in the case of *local* operation invocation.

of all available 8 timeslots of a single GSM channel. Practically, the data rate per direction tends to be limited to about 12 kbit/s for the uplink and about 60 kbit/s for the downlink, depending on the device used and the availability of free channel timeslots in the network. All in all, GPRS is not suitable for time-critical or high-bandwidth communications, but sending commands and receiving status reports via GPRS basically permits true long-range missions with ground-station interaction possible.

At the time of writing, flat-rate contracts for GPRS data transmission are available at about 40 EUR/month (e-plus, Germany). This is clearly well affordable both in research and in application cases.

More and more suitable GPRS modem devices are being introduced into the market. One arbitrary example is the *MultiModem* product line by MultiTech, Inc. [98]. These modems are available with RS 232, USB, Bluetooth, or Ethernet connectivity, and even with an integrated GPS receiver[1]. The RS-232 version weighs 119 g including box.

Of course, GPRS modems need to be equipped with a registered GSM SIM card to be operational. Any Internet-connected PC can serve as the peer communications endpoint – with no Internet connection available, a second GPRS modem may be required at the ground station.

4.3 CORBA

This section briefly introduces the well known and widely used CORBA standard (*Common Object Request Broker Architecture*) developed and maintained by the *Object Management Group* (OMG) and investigates its suitability for use as a "middleware" abstraction layer on board small UAVs. First, an overview of the CORBA core architecture is provided. Then, features within the CORBA framewok are identified and discussed that are particularly relevant to distributed real-time systems.

---

[1]Of course, this is a "natural" product for use in modern highway toll applications, like the *TollCollect* system in Germany.

4.3.1   Core Architecture

The CORBA core in its most recent version is defined in [64], with backward compatibility to prior versions of the specification. CORBA is intended to facilitate the design of distributed applications. Its most central foundations are

1. a client-server paradigm, and

2. *remote procedure call* (RPC) semantics.

According to the client-server paradigm, communication in CORBA is always initiated by a *client* application as the invocation of some method to be performed by, or to, some *object* representing the *server*. As this invocation is performed synchronously, i.e. blocking the client during the operation, and allows the transmission of input and output parameters to and from the operation, it carries the classical semantics of a remote procedure call (RPC). CORBA's extension to the latter consists in the notion of objects: by providing an object reference together with the method invocation, the client can address and manipulate a particular item's state, which may persist between multiple operations. This is fully analogous to the transition from procedure calls to method invocation as resulting from the concept of *object-oriented programming*.

The software component – or software layer – offering the CORBA service as described above is the *object request broker* (ORB). Figure 4.2 depicts the software layers concerned with the invocation of a method provided by an object implementation *on the same machine as the client* ("local" invocation): The client application uses the *stub* code specific to the method to be invoked. The stub is responsible of transparently coding the parameters of the invocation into some transfer syntax (*marshalling*) and making the returned result of the operation available to the application (*unmarshalling*). The *skeleton* code, on the other hand, decodes the parameters to be understandable by the object application, which may include byte-order conversion, and marshals the invocation result, if any. In order to permit the automatic generation of compatible stub and skeleton modules, CORBA requires the objects' invocation interfaces to be specified via the data description language IDL (*interface description language*). IDL is defined within the basic CORBA specification [64]. Using the same IDL specification for compiling both client stubs and object skeletons, application-level software may be written in any programming language for which an IDL mapping has been defined – this is the case for, at least, C, C++, Java, COBOL, Smalltalk, Ada, Lisp, and Python. Application programmers cannot – and need not – find out in what language any peer component has been programmed.

It has to be noted that the interface between the stubs and skeletons on the one hand and the ORB on the other is actually proprietary. Thus, the stub and skeleton IDL compilers have to be specific to the ORB implementation used.

The simple scenario depicted in figure 4.2 is not, however, sufficient to explain the benefits to be obtained from CORBA. These become more obvious when "remote" invocations come into play, with client application and object implementation running on different hosts. This is depicted in figure 4.3: While the stub and skeleton mechanism works as in the local invocation case, there are two ORBs involved. They are running on two different hosts and need to

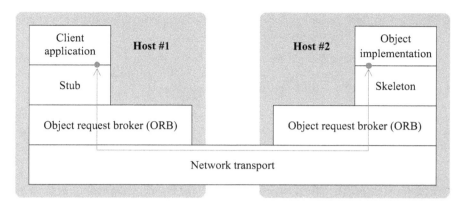

Figure 4.3: CORBA core structure in the case of *remote* operation invocation.

employ some underlying network connection to pass the invocation request and result between the client application's and the object implementation's system. While the object reference is required to tell the client's ORB about the actual location of the object implementation, this is fully transparent to the client: it cannot, and need not, distinguish local and remote invocations based on the object reference.

The CORBA standard does define a communication protocol between ORBs for remote invocations, the *general inter-ORB protocol* (GIOP). A specialization of the GIOP using TCP/IP as the network transport and called *Internet inter-ORB protocol* (IIOP) ensures immediate interoperability between any two ORB implementations compliant to the CORBA standard.

In order to implement the basic scheme explained so far with regard to a wide variety of application-specific requirements, much more complexity has been added to the CORBA standard. The following list only selects a few of the related concepts and keywords:

- While client applications can – and should – always assume unlimited lifetime of the objects they are referring to, implementations must take care of actual server resources. The process or thread performing operations associated with an object is called *servant*. The server-side entity deciding about the activation and deactivation of servants, fully transparent to client applications, is the *portable object adapter* (POA). The POA supports a number of different *policies* for implicit activation and deactivation of servants, which can be selected according to the resource and response-time requirements of the application.

- In addition to the "static" stub-based invocation scheme, ORBs additionally provide the *dynamic invocation interface* (DII). Using the DII, clients can dynamically query the IDL interface definitions of available objects from the *interface repository* (IFR) and dynamically invoke their methods, even if the corresponding interfaces were not known to the programmer. The resulting invocation is then a CORBA object itself.

- ORBs need to offer a *naming service* that permits clients to initially obtain object references to be used in method invocations.

- CORBA provides basic support for *load balancing* on the server side. That is, server-side ORBs may transparently route client requests to different hosts or systems in order to pool great amounts of distributed computing resources.

All in all, it should have become clear that CORBA is primarily designed to fulfill the middleware requirements of very big, complex, and flexible production systems. There is much focus on maximum independence among the system components: through the IDL interface specifications and transparent object references, parts of the system may be physically moved or even implemented in a different programming language without any other component being aware of this change. Therefore, large-scale distributed business accounting, reservation, or other database systems would be among the first application fields imagined.

However, these system properties are rather untypical of embedded real-time systems. In the latter, it is usually necessary to consider *all* system components and their coupling to infer real-time bounds, and the system architecture is relatively simple and fixed. Furthermore, especially with respect to small UAVs, resource limitations in such systems tend to be more stringent so that abandoning flexibility in favor of saved resources will clearly be more attractive in the small UAV case than in most other cases of middleware employment.

Furthermore, explicit synchronous method invocation does not really suit the communication requirements of small UAV on-board computing tasks as depicted by figure 3.1. Here, information exchange is dominated by fixed unidirectional communication streams with state – instead of event – semantics, as already explained in section 4.1. The following subsections deal with extensions and specializations of CORBA addressing this type of applications.

4.3.2   Real-Time Extension

In order to open up the field of real-time systems to the application of CORBA, OMG introduced the first *Real-Time CORBA* extension specification in 2000. Today, there are two concurrent standard versions available, one restricted to static scheduling [67] and one additionally supporting dynamic scheduling [63].

The basic idea behind Real-Time CORBA is to provide an interface and concepts to help including CORBA in a distributed real-time system. It does *not*, however, provide transparent real-time semantics that would allow some application to be executed on top of *any* Real-Time CORBA ORB, meeting all specified real-time bounds whenever there are sufficient computing resources. This means, for example, that fixed priorities (in the static scheduling case) or parameters like deadline and estimated execution time (in the dynamic scheduling case) need to be set by the application.

The most important features provided by the Real-Time CORBA extension are the following:

Priority Mapping provides a transparent conversion between system-wide CORBA priorities and the priority schemes used by different local real-time operating systems. Application software is required to specify priorities solely via the global

`RTCORBA::Priority` type. The actual local client thread priorities are mapped into CORBA priorities upon generation of an invocation request, and re-mapped into server-side local priorities upon method invocation.

Priority Banding refers to the application's ability to reserve communication channels (connections) for certain subsets of CORBA priorities in order to facilitate the appropriate use of available communication bandwidth.

Threadpools enable the server-side software to reserve computing resources (threads) in a static and/or dynamic way in order to reduce the risk of thread starvation.

Invocation Timeouts limit the maximum blocking period of a thread after invoking a method. While returning an exception upon timeout does not really solve the underlying response time problem, it facilitates error handling and may reduce the resulting effect on overall system behavior.

All in all, Real-Time CORBA provides several suitable features to make the underlying real-time operating systems' services available to and manageable by the distributed application. Unfortunately, it does not address the issue of communication bandwidth scheduling. Instead, the client software is expected to limit its rate of method invocation according to the available bandwidth, while the system considers every single invocation having to be eventually satisfied. Regarding the overall complexity of CORBA and its computational requirements, this failure to support state-propagation type communication under real-time conditions is unsatisfactory.

4.3.3 Messaging and Asynchronous Method Invocation

The current CORBA specification [64] introduces additional extensions to the basic services related to *messaging* and *asynchronous method invocation* (AMI) in chapter 22. An ORB's compliance to CORBA Messaging is optional. This extension permits the relaxation of the synchronous semantics of method invocation. For this purpose, the creation of an asynchronous invocation request is seen as the sending of a message, while the arrival of this request at the target object conforms to the reception of the message.

From the viewpoint of ORB implementation, AMI affects the client side only. The server side need not care about the synchronicity of method invocations in general. In order to permit the invocation of any kind of method – including methods with return values and output parameters – in an asynchronous way, AMI defines two different client-side interfaces to handle these invocation results:

Callback Model. This model requires the client to register a separate `ReplyHandler` object that receives the invocation results after completion. The signatures of both the asynchronous invocation operations and the `ReplyHandler` object are automatically generated by a messaging-enabled IDL compiler from the object interface specification.

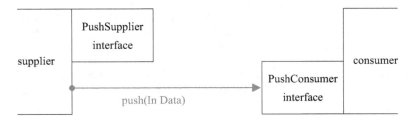

Figure 4.4: CORBA event service using the push interface. The arrow indicates the initiative of invoking the event interface for data transfer.

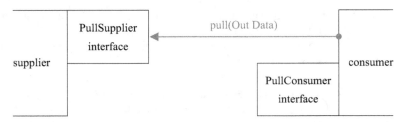

Figure 4.5: CORBA event service using the pull interface. The arrow indicates the initiative of invoking the event interface for data transfer.

Polling Model. This model creates a separate `Poller` object at the client side containing operations that correspond to the remote object's methods and allow the client to explicitly query the return value and output parameters resulting from a previous asynchronous method invocation using the polling model. The signatures of both the asynchronous invocation operations and the `Poller` object are automatically generated by a messaging-enabled IDL compiler from the object interface specification.

While AMI semantics are in many cases better suited for embedded real-time systems like small UAVs than synchronous invocation, their probable client-side implementation adds even more overhead to the use and execution of such invocations. Instead, truly decoupling the request and response aspects would have been preferable regarding resource consumption and development complexity in embedded real-time scenarios.

4.3.4   Event and Notification Services

Another drawback of the CORBA services described so far is the lack of facilities for one-to-many flow of information, e.g. for the distribution of state information between system components. Again, see figure 3.1 for the relevance of this to the kind of systems discussed within this book.

To overcome this limitation, OMG has specified the *Event Service* [65] and the *Notification Service* [66]. Both specifications are outside the CORBA core, but supposed to be realized

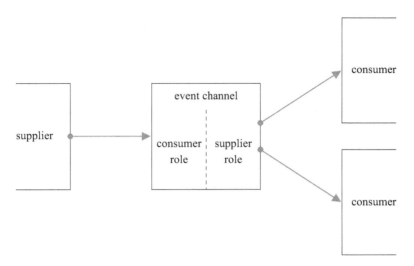

Figure 4.6: CORBA event service using an event channel for one-to-many communication. The event channel may use any combination of push and pull interfaces at the connections (here, the arrows indicate the direction of data transfer).

using the CORBA core services. Furthermore, the notification service is a proper extension of the event service. [13] is a highly informative paper in which the authors discuss the suitability of both services for use in a large distributed real-time application, i.e. the shared distributed commanding of multiple robot systems.

The event service offers two different models that differ in the initiative of data transfer. Each model requires the implementation of a model- and role-specific interface by both the supplier and the consumer of the events in question:

Push Model. In the push model, the supplier initiates the transmission of an event by invoking the push operation of the PushConsumer interface. The PushSupplier interface is only required for disconnection on behalf of the consumer. Figure 4.4 depicts the employment of this model.

Pull Model. In the pull model, the consumer initiates the transmission of an event by invoking the pull operation of the PullSupplier interface. Alternatively, the consumer may invoke try_pull in order to poll the availability of an event at the supplier. The PullConsumer interface is only required for disconnection on behalf of the supplier. Figure 4.5 depicts the employment of this model.

Regarding the generation of events and the associated overhead, the push model seems to constitute the more natural option for use in an embedded real-time system.

In order to overcome the point-to-point nature of communication, the event service specification introduces *event channels*. An event channel – or *proxy* – is a separate entity acting both as consumer and as supplier, storing and relaying events between the original supplier and one or many final consumers. Figure 4.6 shows an example of using an event channel. For every single connection in this setup, any of the two models explained above may be used. The specification introduces additional interfaces for setting up and managing communication relations via event channels.

The notification service adds more flexibility to the event service. The most relevant features introduced by the notification service include:

- The ability for clients to attach *filters* to proxy objects. Thus, every registered consumer can decide to only receive a subset of the events issued by the supplier.

- The ability for suppliers to find out what event types are actually received by consumers, in order to prevent the generation of irrelevant events.

- The ability for consumers to find out which events are currently offered by suppliers, so that they can react to new event types becoming available.

- The ability to assign quality-of-service (QoS) attributes to events and channels, so that the utilization of communication resources and the contention between different events can be explicitly controlled to some degree.

As worked out in [13], the notification service finally fulfills the basic communication requirements found in large distributed real-time systems quite well. By limiting the number of queued events via QoS management, it is even possible to acquire a good approximation to state semantics.

Yet, it needs to be noted that to use these features, a full CORBA implementation is required, plus an implementation of the notification service running on top of it. The notification channel proxies involved in fulfilling the notification service are, in general, CORBA clients themselves, competing for computing resources with the "original" components of the real-time system. Furthermore, not even the notification service addresses the issue of bandwidth scheduling in a transparent way: while clients may exert some influence on the latter, global real-time guarantees on communication would still require the cooperation – and consideration – of all client components that generate events.

4.3.5   Minimum CORBA

In the consciousness that the resource requirements of the full CORBA core may threaten its applicability in certain contexts, OMG has issued the *minimumCORBA* subset specification [62]. It is aimed at retaining the full interoperability of most CORBA applications while disposing of expensive features that are rarely used. The omissions defined by minimumCORBA include:

- The **dynamic invocation interface (DII)** is not supported, because it requires a complex interface and is unnecessary in normal (static) application design.

- **Policies** for automatic servant activation and deactivation are not supported to reduce the degree of dynamics on the server side.

- Several features deprecated in the current core specification are omitted.

The minimumCORBA subset permits the implementation of the notification service. This infrastructure would constitute the most attractive scenario of using CORBA in an embedded real-time system. Yet, minimumCORBA and notification service together would still exhibit an considerable footprint.

4.3.6 Subsumption

After discussing the various features and variants of CORBA in the previous sections, this section subsumes the suitability of CORBA for use in small UAV on-board systems:

- CORBA, possibly minimumCORBA, together with OMG's notification service, support many, but not all communication requirements met in a small UAV on-board system. The main desirable but missing feature is the transparent scheduling of inter-component communication bandwidth.

- CORBA would impose considerable redundant system complexity in both the programming and the execution phases.

- The resulting consumption of memory and computing time resources due to the use of CORBA might be acceptable in a large ground-based distributed real-time system, especially when the individual nodes have to run a full-scale real-time operating system. However, weight and energy constraints imposed by small UAV on-board use do render these resource requirements maximally undesirable.

All in all, CORBA's only partial fulfillment of the communication requirements should fail to pay for its footprint and complexity here.

4.4 BBCS

This section introduces and describes the *Blackboard Communication System*, BBCS, a middleware system meeting a superset of above requirements. Certain aspects of BBCS have already been described elsewhere, e.g. [113, 104]. This presentation specifically addresses the bandwidth scheduling algorithm used by BBCS formally and depicts the usage of BBCS for state variable transmission.

The subsequent section 4.4.1 summarizes all features provided by BBCS. Section 4.4.2 covers BBCS's algorithm for bandwidth scheduling and the resulting real-time guarantees. Section 4.4.3 explains the application of BBCS to state variable transmission by means of examples.

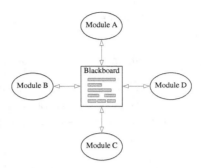

Figure 4.7: Blackboard architecture view of BBCS.

4.4.1   Features

This section gives a brief overview of the design features implemented in the BBCS middleware.

Architecture

A distributed system utilizing BBCS consists of *nodes* and point-to-point communication *channels*. Neglecting broadcast-type communication media in modeling does not impose any unsuitable restriction, because

1. broadcast-type communication media can still be used to establish multiple point-to-point channels, and

2. broadcast-type communication modeling would have been fairly undesirable due to the resulting medium-wide scheduling constraints.

The underlying *blackboard* architecture denominates some virtual memory storage that is transparently synchronized between the BBCS nodes via the BBCS protocol (see figure 4.7). With respect to state-variable type data, this provides almost complete abstraction from distributedness.

BBCS Protocol

Every BBCS channel fulfills the requirements listed in section 4.1. In addition to supporting state-variable class communication, BBCS also offers a service to transmit reliable data streams with flow-control within the channel's data traffic. The distinguishable data portions are called *slots* and indicated in figure 4.7 through the grey rectangles. Each slot thus constitutes either a certain state variable or a certain data stream and can be assigned a fixed guaranteed fraction of a channel's bandwidth. The BBCS protocol subdivides the exchange of data into fixed-size packets, called *stripes*. The size of these stripes, a configuration option, defines the minimum granularity of bandwidth scheduling. State variables may occupy much more memory than the stripes' size. Then, BBCS will use as many stripes as required to transmit a new content of the state variable, but the update will be presented as atomic to the user in order to preserve the non-distributed abstraction. For every slot's contents, there can be at most one node in the network acting as the *data source*, but as many nodes as desired may read the slot's data (*data sinks*).

Shortest-Path Routing

In a BBCS network with many nodes and channels, there can be more than one possible transmission path between a data source node and a corresponding sink node. BBCS performs dynamic shortest-path routing throughout the network. Dynamic here means that an alternative route is set up automatically if one of the channels suffers a temporary failure and an alternative route of working channels still exists. The initial shorter route will be re-enabled as soon as the faulty channel resumes its operation.

The routing algorithm is based on a classical *distance vector routing* approach: Every node records, for every slot, the minimum number of point-to-point hops with which it may receive the respective slot's contents, and reports this distance vector to its neighboring nodes. From a global perspective, the distance vectors locally computed from the neighboring nodes' results converge to correct solutions within a time span linear in the length of the longest transmission path in the graph of nodes. Any nodes requiring the reception of the contents of a particular slot requests these data from a single peer node, which it selects according to the local distance vector.

Platform Independence

In order to achieve maximum independence of operating systems and communication devices, the BBCS's design has been split into a big platform- and device-independent main part and a very small platform-dependent part. The latter is called *platform abstraction layer* (PAL), providing only a very small set of services (basically the transmission and reception of raw data packets, memory management, and real-time clock). So far, PALs have been successfully implemented for and BBCS utilized on the following set of platforms: Solaris, Linux, Windows, Windows CE, generic POSIX, QNX, Infineon SAB80C166/167 microcontroller family, Atmel ATmega128 microcontroller, and LabWindows CVI real-time controllers. Low-level communication devices and protocols supported by the PAL implementations so far are: TCP, UDP, and RS 232 serial links.

Practical Use

BBCS was developed and served as the primary middleware in the European Commission IST project COMETS [28] and for TU Berlin's UAV MARVIN [68, 102].

4.4.2 Bandwidth Scheduling

This section presents the bandwidth scheduling scheme and algorithm used by BBCS.

As explained in section 4.4.1, bandwidth scheduling is performed for each sending node and channel with *stripe* granularity. In order to guarantee deterministic bounds, a *static* transmission schedule is generated whenever the set of bandwidth requests for the node and channel in question changes. This static schedule consists of a list of 2^K slot numbers $\langle s_i \rangle$, $0 \leq i \leq 2^K - 1$, defining the order of slots to be allowed to transmit exactly one stripe packet. Whenever the end of the schedule is reached, it is applied again from its beginning, and so on. Thus, the effective resolution of the schedule, in terms of the minimum bandwidth fraction $b_{\min}$ to be assigned to a slot, is determined by the parameter K as:

$$b_{\min} = 2^{-K} \tag{4.1}$$

The next section describes the basic algorithm used to process bandwidth requests, called *static binary tree bandwidth scheduling* (SBTBS), and proves its correctness. The subsequent section depicts the obtaining of the $\langle s_i \rangle$ schedule from SBTBS's output. Finally, the resulting guarantees on the transmission delay are pointed out and proven.

Scheduling Algorithm

The SBTBS algorithm constructs a binary tree T the leaf nodes of which are marked with a slot number or the special tag *EMPTY*. Initially, T consists only of its root node, marked *EMPTY*. Every node represents $1/2$ of the bandwidth portion represented by its parent, while the root node represents 100 % of the bandwidth available.

SBTBS further utilizes a list $\langle f_i \rangle$, $0 \leq i \leq K$, of pointers to nodes of T. f_i points to the only empty node at level i of T, or is *NIL* if there is no such empty node. Initially:

$$f_0 := \text{root}(T) \tag{4.2}$$
$$f_j := NIL \quad (1 \leq j \leq K) \tag{4.3}$$

Below, SBTBS is presented for the reservation of a bandwidth fraction of 2^{-i} for slot s, with $0 \leq i \leq K$. 2^{-K} is the resolution of the schedule and constitutes the smallest fraction reservable. Fractions that are no powers of two can obviously be reserved by simply using the same procedure repeatedly for every 1-bit within the fraction's binary representation.

1. Determine the smallest empty node n in T big enough for the requested fraction:

$$k := \max\{j | i \geq j \geq 0 \land f_j \neq NIL\}$$
$$n := f_k \tag{4.4}$$

   If no such k can be found, more than 100 % of the total bandwidth is being requested. In this case, return with failure.

2. Forget the corresponding empty-node pointer:

$$f_k \quad := \quad NIL \qquad\qquad\qquad (4.5)$$

3. While $k < i$ (i.e. while node n represents more relative bandwidth than requested), split its bandwidth fraction among two new empty child nodes, recording the "right-hand" child in the list $\langle f_i \rangle$, and proceed with the "left-hand" child:

$$
\begin{aligned}
k \quad &:= \quad k+1 \\
m \quad &:= \quad \text{add\_child}(n, EMPTY) \\
f_k \quad &:= \quad \text{add\_child}(n, EMPTY) \\
n \quad &:= \quad m \qquad\qquad\qquad (4.6)
\end{aligned}
$$

4. Now, n is an empty node with the correct size. Mark n with slot number s. Return with success.

The correctness of SBTBS can be proven by showing two propositions:

Proposition 1 *Between any two requests, the nodes listed via $\langle f_i \rangle$ together exactly represent the total bandwidth fraction not yet reserved.*

Proposition 2 *SBTBS completes successfully iff a sequence of bandwidth requests does not exceed 100 %.*

Proof of proposition 1: By induction over the number of requests. Initially, the root node is tagged *EMPTY*, pointed to by f_0, and represents 100 % of the available bandwidth. This anchors the induction. Induction step over one execution of SBTBS: Node f_k was initially empty because of step 1 and is removed from $\langle f_i \rangle$ in step 2. For each execution of the *while* body in step 3, the remaining portion such removed from $\langle f_i \rangle$ and not yet assigned is split in two, one half newly included into $\langle f_i \rangle$ and the other half kept recorded in n, also a portion unassigned but not currently present in $\langle f_i \rangle$. Here, it is important to note that all pointers f_j, $k+1 \le j \le i$, with k from step 1, were *NIL* according to the calculation of k. Finally, the node referred to by n is assigned to the requesting slot, thus correctly no longer listed in $\langle f_i \rangle$. $\square$

Proof of proposition 2: Whenever a "suitably big" free node is found in step 1, SBTBS succeeds. This is correct because the node found in this way was unused, by prop. 1. Otherwise, by prop. 1 and step 1, all single remaining free nodes are smaller than the request 2^{-i}, which means that their sum is also smaller than 2^{-i}, because:

$$\sum_{j=i+1}^{K} 2^{-j} \quad < \quad \sum_{j=i+1}^{\infty} 2^{-j} = 2^{-i} \qquad\qquad\qquad (4.7)$$

Therefore, if step 1 returns failure, the request must have exceeded the total remaining bandwidth portion. $\square$

Figure 4.8 visualizes the execution of SBTBS for the following request list, from left to right then top to bottom:

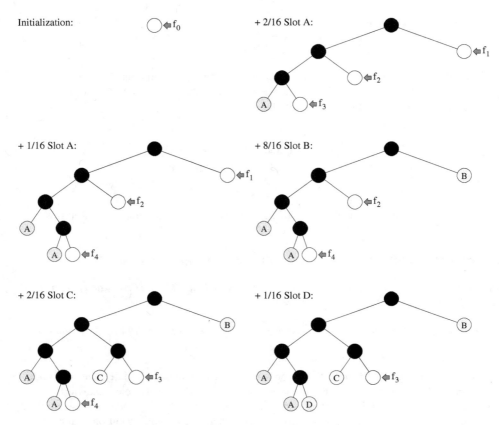

Figure 4.8: Example run of the SBTBS algorithm, $K = 4$, $b_{\min} = 1/16$.

Slot	A	B	C	D
Bandwidth request	3/16	8/16	2/16	1/16

Please note that slot A is processed through two subsequent requests, representing $3/16 = 2 \cdot b_{\min} + 1 \cdot b_{\min}$. The arrows depict the list $\langle f_i \rangle$ in every state. Finally, one empty node representing the unassigned fraction of 2/16 is left.

Derivation of Schedule

It remains to be seen how the schedule in the $\langle s_i \rangle$ representation is obtained from the final tree T. It is essential to spread every bandwidth assignment over the schedule as evenly as possible, while the nodes of T show the most compact distribution – which is in fact just the opposite extreme. The "most even" distribution is easily obtained by:

1. Labeling the two outgoing edges of every node with 0 and 1, respectively.

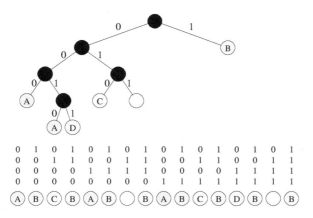

Figure 4.9: Derivation of schedule s_i from the SBTBS output.

2. Setting s_i according to the tag of the leaf node that is reached from the root, by following the edges as determined by the binary representation of i, from least to most significant bit (LSB to MSB).

Figure 4.9 depicts this for the tree from figure 4.8. The "empty" entries s_6, s_{14} will be skipped on executing the schedule, in favor of the effective bandwidth provided to each of the slots.

Real-Time Bounds

Proposition 3 *Let $d_{\max}$ denote the maximum delay possible, in stripe transmissions, between two subsequent transmission times of a stripe of slot n with an assigned bandwidth fraction r, according to the schedule generated above. Then:*

$$d_{\max} \;=\; 2^{-\lfloor \mathrm{ld}\, r \rfloor} \tag{4.8}$$

Proof of proposition 3: For a total relative bandwidth r, the biggest fraction of r assigned through a single call of SBTBS is 2^b, with $b = \lfloor \mathrm{ld}\, r \rfloor$ denoting the exponent of the most significant bit of the binary representation of r. Therefore, the path to the corresponding node of T is $-b$ edges long, and has thus been periodically selected every 2^{-b} positions during the construction of $\langle s_i \rangle$. This already implies (4.8). $\square$

Please note that this $d_{\max}$ is equal to the theoretical optimum of $1/r$ for proper powers of two, $r = 2^{-b}$, and always better than the theoretical optimum doubled:

$$1/r \;\leq\; d_{\max} \;<\; 2/r \tag{4.9}$$

4.4.3 Usage

This section provides an overview of using BBCS in an application by means of examples. These examples deal with state-variable type communication, reliable data streams are not addressed here (their usage does not significantly differ from usual `read` and `write` operations).

The following two subsections deal with different example scenarios, transparent periodic access on the one hand and occasional data push on the other. All examples are basically written in C. Portions regarding the initialization of BBCS and the communication channels are not considered.

Transparent Periodic Access

Transparent periodic access means that the application can use a state variable just as if it were fully local. A convenient example for this kind of situation would be a simple controller that gets the current speed vector $v \in \mathbb{R}^3$ of some UAV as its input and adjusts the UAV's elevators in order to keep the vertical speed (say, v_2) close to zero. The input and output variables `vel` and `elv` shall be defined, in fixed-point notation, in this way:

```
long  vel[3];
long  elv;
```

Here, it can be assumed that `vel` is measured and `elv` applied by another system node than the controller is running on.

Now, the usage of BBCS within the controller node mainly consists of:

1. Some initialization code, binding a BBCS slot to the two "blackboard" variables:

   ```
   bb_add_slot(VEL_SLOT, 0, sizeof(vel), &vel, 90);
   bb_add_slot(ELV_SLOT, 0, sizeof(elv), &elv, 30);
   bb_auto_get(VEL_SLOT);
   bb_auto_put(ELV_SLOT);
   ```

   Here, VEL_SLOT and ELV_SLOT are just constants to identify the slots. 0 is the flag defining the two slots declared here to use state-variable semantics (as opposed to reliable data-stream mode). The size and address expressions bind the user variables to the corresponding slots, and 90 and 30 are the bandwidth fractions (in 1/1000) to be assigned to the slots. Note that the `vel` variable occupies more memory than `elv`, so it is reasonable to assign bandwidth correspondingly.

   `bb_auto_get` and `bb_auto_put` instruct BBCS to transparently synchronize the slots' user variables whenever an update is received, or whenever bandwidth is available, respectively.

2. The control loop, where there is no trace of distributedness:

   ```
   elv = -42.17 * v[2];
   ```

(This is a simple P controller with some arbitrary gain, but this detail does not matter here.)

3. Somewhere in the main loop of the controller process, the instruction to BBCS to perform all pending synchronization:

```
bb_sync();
```

It has been a deliberate design goal that BBCS operates synchronously to the user application, requiring bb_sync for the explicit invocation of synchronization activities. This simplifies user applications considerably.

Occasional Data Push

Occasional data push means that updates of a slot variable are relatively seldom, and the update events need to be handled explicitly. A possible example would be the changing of a camera's focal length, particularly if the corresponding zooming operation requires significant extra time. Using a variable

```
int f;
```

for the focal length in mm, the code might look like:

1. For the initialization, assigning 1 % of the available bandwidth:

```
bb_add_slot(F_SLOT, 0, sizeof(f), &f, 10);
```

2. At the *sending node*, a human user might trigger the change of focal length, resulting in a function call to:

```
void changeFocalLength(int value)
{
  f = value;
  bb_put(F_SLOT);
}
```

At the *receiving node*, the resulting slot updates can be detected in the camera control loop like this:

```
if (bb_get(F_SLOT) > 0) {
  setCameraFocalLength(f);
}
```

3. Also in this case, both nodes must periodically perform:

```
bb_sync();
```

4.5   Subsumption

This chapter has discussed software and hardware required for communications within small UAV systems. Specifically, aspects covered have been middleware and communication devices.

Off-the-shelf middleware is widely unsuitable for this kind of application. This situation might improve in the future, but today, there is good reason for employing a dedicated middleware layer, which may have to be specifically designed.

With regard to communication devices, it must be noted that basically all available technologies either suffer from undesirable range limitations, or rely on public infrastructure, which in turn may compromise their reliability and suitability for safety-critical communication tasks. Therefore, it will often be desirable to design small UAV system in a way that depends as little as possible on communications for safety reasons, so that the requirements to communication technology may be as far as possible relaxed.

Chapter 5

Flight Control

Flight control for stabilization and trajectory flight of a small UAV may seem to constitute the most challenging task in the design of the system architecture of a small UAV. Indeed, there is a great amount of recent literature about nonlinear, adaptive, underactuated control systems and their relevance to UAV flight control. This literature may be read as implying that these problems related to UAV flight control are particularly difficult to solve – which, from a practical point of view, is not at all true.

This chapter discusses requirements and possible solutions associated with the flight control task. After the usual introductory section 5.1 about requirements, section 5.2 gives a very condensed overview of the methods of UAV flight control recently addressed in control systems literature. Section 5.3 then presents the formal control system design underlying the flight control of the MARVIN UAV, which is based on cascades of *backstepping controllers*. The following section 5.4 performs a closer investigation of trajectory control of multi-integrator cascades. These are, on the one hand, merely a special case of the cascades discussed in section 5.3, but comprise, on the other hand, a very handy view on motion control as encountered within every UAV system and do exhibit some formal beauty. A by-product of this view is a simple way of multi-point trajectory tracking. After these fundamental sections, sections 5.5 to 5.7 present examples of complete UAV flight control architectures based on the formal work developed in this chapter. Section 5.8 finalizes this chapter with a critical subsumption of its findings.

It should be noted that this book does not constitute a control theory textbook. Therefore, there will be no emphasis on maximum generality, but on the requirements of small UAV control and on successful yet manageable design procedures of small UAV flight controllers. The described approach to controller design is pretty *straightforward* – in the sense of modularity and intelligibility more than in the sense of being identical to the methods taught in traditional basic courses of control theory. This book is focussed on a single application field, not on bringing forth new mathematical results in control theory.

Still, *global asymptotic stability* (GAS) of the resulting convergence solutions will be formally guaranteed under certain reasonable assumptions about the available system model.

5.1 Requirements

The following list of items addresses the requirements that must typically be met by small UAV flight controllers and some intrinsic implication of small size to flight dynamics:

- Small UAVs, like all aerial vehicles, constitute *underactuated* systems: The state space spanned by the model of an aerial vehicle usually comprises 12 scalar state variables at the least. Namely, there are three *degrees of freedom* (DOF) in each of position, change rate of position, orientation, and change rate of orientation. On the other hand, the number of control inputs rarely exceeds 4 or 5. Therefore, the desired trajectory in the state space needs to be stripped down to as many DOF as there are control inputs. These control inputs then need to be applied such that the desired trajectory is followed, implicitly using the remaining state variables suitably to fulfill this purpose.

- UAV dynamics exhibit a certain degree of nonlinearity. Although this fact alone may already discourage the application of a large set of control methods, it is important to note that there are basically two different kinds of nonlinearities that should be considered independently:

  1. The dynamic response to some control input may be nonlinear from the start. This applies to, e.g., aerodynamic equations that correlate force with squared velocity, or may result when two different effects combine, like angular momentum and aerodynamic drag of a propeller.

  2. *Coordinate transforms*, e.g. between world and body frame, are always nonlinear due to their dependence on the current orientation angles of the UAVs. While the nonlinearities resulting from coordinate transforms necessarily precipitate in the full dynamics model of the system, there may still be perfect linear correspondences in a properly chosen coordinate system.

  Whenever most of the nonlinearities have been obtained through coordinate transforms, which is very frequent in UAV modelling, the art of control system design consists in carefully addressing every dynamic subsystem most easily and effectively in the corresponding coordinate system, getting fully rid of the adverse side-effects of coordinate transforms.

- Due to the effect of wind and aerodynamic turbulence, UAV flight controllers usually have to cope with a significant level of unpredictable disturbances. This also means that closed-loop stability proven according to some simplified system model needs to be interpreted with a particular amount of care.

- Generally, the smaller an aerial vehicle is, the faster its dynamics are. This imposes a special challenge on the sensor data processing rate and the rate of computed control cycles in the small UAV context. This general fact shall be explained by means of an example: Consider the angular acceleration $\dot{\omega}$ incurred by an aerial vehicle of characteristic size (radius) r, caused by some external aerodynamic disturbance. Writing down

the dynamics laws with all constants dropped, it is:

$$\dot{\omega} \quad \sim \quad \frac{M}{J} \quad \sim \quad \frac{Fr}{mr^2} \quad \sim \quad \frac{r^2 r}{r^3 r^2} \quad \sim \quad \frac{1}{r^2} \tag{5.1}$$

Here, M is the torque acting on the vehicle, J is its moment of inertia, m the vehicle's mass (assuming mass distribution and material independent of the vehicle's size). Obviously, the vehicle's agility responding to external aerodynamic disturbances scales with the "square of its smallness". The principal reason for this drastic effect is the proportionality of the moment of inertia, through the combination of radius r and mass m, to r^5.

5.2 Approaches Reported in the Literature

This section provides a short and selected overview of control engineering approaches proposed in the literature for application to UAV flight control. Only approaches successfully applied for flight control in practical flight experiments have been considered.

5.2.1 Monolithic Linearization

A rather traditional approach is the monolithic identification of a linear system approximation, as performed e.g. by Shim, Sastry, and others in [125] for helicopter flight control. Here, the system is modeled through a linear equation

$$\dot{x}(t) \quad = \quad A \cdot x(t) + B \cdot u(t) \tag{5.2}$$

with $x \in \mathbb{R}^n$ denoting the system state, $u \in \mathbb{R}^m$ the control input, and A and B matrices describing system behavior. This model can then be used as a basis for classical linear control system design. This approach has two significant drawbacks:

1. System identification can be tedious, as not all state variables may be observable. Then, numeric estimation procedures may have to be employed, possibly involving issues of numerical stability themselves [125].

2. Modeled monolithically, UAV dynamics are always nonlinear, due to rotational coordinate transforms at least. (5.2) can only capture a local linearization at a specific operating point, which may be of varying utility throughout the full state space of the system.

Nevertheless, even this approach is suitable for the successful design of a high-performance helicopter flight controller for near-hover conditions, as reported in [125, 126].

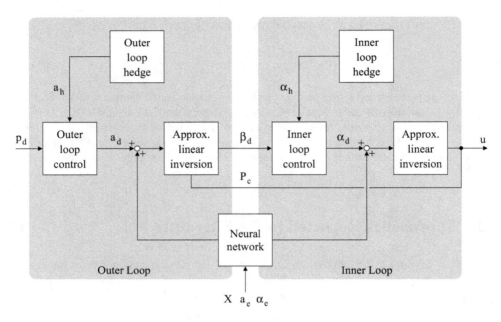

Figure 5.1: Schematic overview of a dual-loop controller with neural adaptation and pseudo-control hedging.

5.2.2   Nonlinear Model Predictive Tracking Control

In [126], Shim, Sastry, and others report the very good performance of *nonlinear model predictive tracking control* (NMPTC) when applied to small-scale helicopter flight control. In this approach, the trajectory tracking error and control inputs required are predicted for a limited time horizon using a nonlinear system model and evaluated by means of a quadratic cost function J. At controller runtime, the predicted cost $J(t)$ is numerically minimized at every time step using gradient descent. This controller results in very good performance even in the presence of model errors and is reported by the authors to be clearly superior to simple monolithic linear control as of 5.2.1 above.

However, examining the approaches compared in [126] and the method proposed in the remainder of this chapter, it seems that the superiority of the NMPTC approach is mainly due to the weakness of the linear controller it is compared with. Furthermore, the authors attribute a "tremendous computing load" [86] to the NMPTC controller and suggest the execution on a separate high-performance on-board computer outside of hard real-time requirements [126].

5.2.3 Inner-Outer-Loop Separation with Neural Adaptation and Pseudocontrol Hedging

The Georgia Tech group suggests to use an approach with two separate control loops with model adaptation through a neural network and *pseudocontrol hedging* (PCH) for helicopter UAV control. [79] provides a current and in-depth coverage of this method and its performance.

This approach employs a hierarchical control system consisting of an *outer loop* for position control and an *inner loop* for attitude control. The outer loop provides the reference signal for the inner loop and uses the latter as it were a simple actuator.

Figure 5.1 gives a simplified overview of this controller. The outer loop receives the desired position signal p_d, from which the outer loop control law calculates a desired acceleration a_d in world coordinates. Subsequently, an approximate linear model is used in its inverted form in order to obtain the helicopter's orientation vector β_d and collective pitch P_c required to cause the desired acceleration. β_d is the reference input to the inner loop, which is structurally analog to the outer one and calculates the remaining control inputs to establish the desired angular acceleration α_d. The final control input vector u effectively consists of the collective and cyclic main rotor pitches and the tail rotor pitch.

As the inverted dynamics models are only linear approximations, a neural network is used to calculate offsets compensating for the model errors in both loops. The network is the *adaptive element* in the proposed controller. According to [79], it may be a multilayer perceptron with a single hidden layer. The network receives the system state X and estimated linear and angular accelerations a_e, α_e according to the simplified dynamics model as inputs and is trained using a gradient-descent error minimization procedure during the operation of the controller. The "training set" for the network is constituted by the current deviation between desired and actual system responses.

The two *hedge signals* a_h and α_h make up for the fact that the system's response to the virtual and actual control inputs is not immediate, but subject to inner loop and actuator dynamics and (possibly) range and rate saturation. The error resulting from these dynamics needs to be eliminated from the neural network's training data, for it cannot be compensated anyway and would interfere with the desired adaptation task of the network. Again, simplified dynamics models are used to predict the "hedging error". This way of restricting the adaptive process to actual model errors is called *pseudocontrol hedging*.

The structure of this controller maps intriguingly well to individual aspects of the design approach developed in the remainder of this chapter. However, it is more specialized to a certain control task, more monolithic, and therefore more difficult to analyze than the iterative approach presented here. Additionally, it suffers from the usual issues with the application of neural networks, which in this case relate to the difficulty in predicting the actual quantitative convergence rate, the possible convergence to a suboptimal local minimum only, and the dependency of the training result on the training set statistics. Accordingly, the authors prove the boundedness of the closed-loop behavior of the neural network and PCH approach, but not its asymptotic convergence.

5.2.4 H_∞ Control

H_∞ control is an approach to control system design with a very strong mathematical foundation. The ETH Zürich group advertises its application to small helicopter flight control e.g. in
[130], and the commercial autopilot system *wePilot1000* [138] uses it as well.

Essentially, the H_∞ norm of a dynamic system is the maximum amplification over the frequency domain that the system can make to its input signal, and H_∞ control uses that one
controller minimizing this norm for the closed-loop system.

[130] reports good results of the H_∞ approach. However, it is a method from linear system
theory, so that its monolithic application to a nonlinear plant, like helicopter dynamics with
nonlinear coordinate transforms involved, is somewhat biased. Moreover, H_∞ control does not
address model adaptation in any special way. Its main attraction is its mathematically defined
optimality.

At this point, it is important to note that mathematical optimality may be significantly different
from the most desirable system behavior from a technical or intuitive point of view. It is a
truly challenging – and rarely attempted – task to accommodate both notions of "optimality"
to one another. Minimum wear, freedom of overshooting, or low strain to human supervisors
are potentially desirable criteria that might be "surprisingly" missed out when targeting at a
purely mathematical optimum.

5.2.5 Critical Remark

At this point, some general critical statement towards the use and role of models in some UAV
control publications is given.

Numerous publications in this field can be characterized by the fact that they use the same
model in the step of controller construction and in the step of performance verification through
simulation. This is somewhat surprising, because the successful verification of a controller
based on the very same model the controller has been designed according to is always a
trivial result. Devising an analogon in physics, this method of working would correspond to
a physicist who derives a theory from an experiment and then predicts the data of the very
same experiment from this new theory. Obviously, this last step can neither fail nor provide
any deeper insight.

Instead, one should either use a considerably less simplified model for simulation than for
the design of the controller, or report the application of the controller to a *real* system in the
results section, or else fully abstain from reporting simulation results as an alleged means of
verification.

5.3   Control Cascade Backstepping

This section develops the methodical basis of the controller design used in the MARVIN
system and its formal stability results.

The core of the suggested design process consists in *integrator backstepping*. This is, in itself, a relatively new but by now widely used technique in higher order nonlinear control design. It is explained, for example, in [88], dating the origins of integrator backstepping, which are partly implicit, around 1988. As the meaning of "backstepping" will become clearest through the formal description of the design process below and due to the lack of a general formal definition of this concept, an informal description suitable for this book shall suffice at this point: Given a dynamic system of second or higher order, backstepping refers to the introduction of one or more *virtual controls* of intermediate (virtual or actual) state variables, *between* the to-be-controlled output x and the actual control input u. The use of these virtual controls gives rise to recursive design procedures of nonlinear control systems. One such procedure is introduced throughout this chapter.

In the broad majority of current literature about backstepping control of nonlinear systems, controller design is strongly coupled with Lyapunov stability theory (e.g. [85]). While the latter provides very powerful tools to ensure stability during the design of controllers for even the most adverse kinds of systems, this book will create an exception from above pattern and do without Lyapunov stability. The reason for deciding against the usual pattern is twofold:

- Lyapunov stability requires the selection of *Lyapunov functions*, which define something like a potential field over the state space and assure system evolution toward minimum potential. But the potential field does not correspond one to one to the employed control law, thus introducing some degree of *redundant* freedom into the design procedure. Hence, it seems justified to avoid this redundancy when easily possible.

- It may interest some readers to learn about the independence of nonlinear backstepping control and Lyapunov functions from the example of this treatment.

5.3.1 Stability Concepts

Since stability analysis will be performed without Lyapunov stability theory throughout this book, the stability concepts to be examined in this book are formalized at this point for reference purposes.

Definition 1 *Given a dynamic system*

$$\dot{x} = f(x,t) \tag{5.3}$$

*with $x \in \mathbb{R}^n$, the point $x = 0$ is **globally uniformly asymptotically stable (GAS)** iff*

$$\forall x(t_0) \in \mathbb{R}^n, \, t \geq t_0 \quad : \quad \lim_{t \to \infty} x(t) = 0 \tag{5.4}$$

Definition 2 *Given a dynamic system*

$$\dot{x} = f(x,t) \tag{5.5}$$

*with $x \in \mathbb{R}^n$, the point $x = 0$ is **globally uniformly input-output stable (GIOS)** iff*

$$\forall x(t_0) \in \mathbb{R}^n \, \exists \varepsilon \in \mathbb{R}_+ \, \forall t \geq t_0 \quad : \quad \|x(t)\| < \varepsilon \tag{5.6}$$

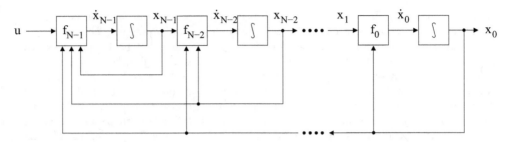

Figure 5.2: Block representation of the class of dynamic systems addressed in this chapter.

5.3.2   Class of Systems

The class of systems to be examined in this chapter is formally defined here. This class is probably broad enough to cover any UAV flight control problem in a way sufficient for practical application. This class consists of all *block-feed-back* dynamic systems

$$
\begin{aligned}
\dot{x}_0 &= f_0(x_1,x_0,t) \\
\dot{x}_1 &= f_1(x_2,x_1,x_0,t) \\
&\cdots \\
\dot{x}_{N-2} &= f_{N-2}(x_{N-1},x_{N-2},...,x_0,t) \\
\dot{x}_{N-1} &= f_{N-1}(u,x_{N-1},...,x_0,t)
\end{aligned}
\tag{5.7}
$$

with

$$
x_j \in \mathbb{R}^{n_j} \tag{5.8}
$$

for $0 \le j \le N$, denoting $u = x_N$ for compactness, and all f_j, $0 \le j \le N-1$, being *continuous* and *invertible* in the sense that there exist all f_j^{-1} such that

$$
f_j\left(f_j^{-1}(\dot{x}_j, x_j,...,x_0,t),\, x_j,...,x_0,t\right) = \dot{x}_j \tag{5.9}
$$

for all states $x_j,...,x_0$ and for all $\dot{x}_j$.

Figure 5.2 depicts this class of dynamic systems by means of block graphics. The block feedback characteristics refer to the fact that only "later" state variables (along the chain of integrators) are allowed as arguments of the (possible nonlinear) transfer functions f_j. This restriction is easily met in most relevant cases. More important a restriction is the invertibility of the transfer functions with regard to their "principal" arguments. Sufficient for fulfilling this restriction is all transfer functions being representable as

$$
f_j(x_{j+1},x_j,...,x_0) = g_j(x_{j+1}) + h_j(x_j,...,x_0) \tag{5.10}
$$

with g_j invertible, which includes many polynomials in x and all transformations affine in x_{j+1} for any given state of $x_j,...,x_0$. Thus, it will be fulfilled anyway in the majority of application cases.

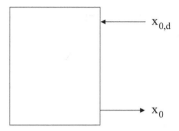

Figure 5.3: Graphical outline of proposition 4.

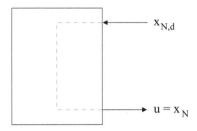

Figure 5.4: Inductive basis for the proof of proposition 4.

5.3.3   Cascade Control Induction

In this section, the controller design procedure proposed in this book is introduced. This is primarily done in the course of an inductive proof of the following proposition 4.

Proposition 4 *Given a dynamic system according to (5.7) and a desired trajectory $x_{0,d}(t)$ continuously differentiable at least N times, any solution $z(t)$ of any[1] closed-loop system*

$$z(t) \quad = \quad x_0(t) - x_{0,d}(t) \tag{5.11}$$

constructed according to figures 5.5 and 5.6 is GAS.

Figure 5.3 depicts the basic idea behind proposition 4 in graphical block notation. The box on the left side represents the dynamic system (plant) according to (5.7) together with a controller. x_0 is the "output" variable of the plant and a corresponding *desired* output signal $x_{0,d}$ as the input to the controller.

Proof of proposition 4: By induction over $x_N, ..., x_0$.

Inductive basis: For $x_N = u$, the "desired output" just equals the control input. Thus,

$$z(t) \quad = \quad x_N(t) - u(t) = 0, \tag{5.12}$$

---

[1]There are infinitely many such systems because of the choice of control parameters.

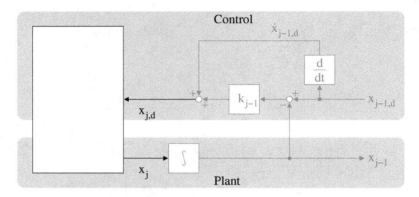

Figure 5.5: Inductive step over a single integrator stage.

which implies $z(t)$ being GAS. Figure 5.4 visualizes this situation.

Inductive step: The inductive step in this proof is performed individually for the integrator stages and the f_j nonlinear transform stages. In this sense, the design procedure corresponding to the proof is technically more general than the system description by figure 5.2, while any sequence of integrator and transform stages could still be covered by the system description after combining transform functions and/or inserting identity transforms. Thus, the proof does cover the same set of systems as the proposition.

- Inductive step over an integrator stage of figure 5.2: Assuming x_j asymptotically tracks the desired value $x_{j,d}$ (inductive hypothesis), consider an augmented closed-loop system with an additional integrator stage in the "plant" part and a linear controller in the "control" part, as depicted by figure 5.5. The control is defined by

$$x_{j,d}(t) \;=\; k_{j-1}\left(x_{j-1,d}(t)-x_{j-1}(t)\right)^{\mathrm{T}} + \dot{x}_{j-1,d}(t) \qquad (5.13)$$

with some gain vector $k_{j-1} \in \mathbb{R}_+^{n_{j-1}}$. Abbreviating the control errors

$$z(t) \;=\; x_{j-1,d}(t) - x_{j-1}(t) \qquad (5.14)$$
$$\tilde{z}(t) \;=\; x_{j,d}(t) - x_j(t) \qquad (5.15)$$

and the smallest component k of the gain vector

$$k \;=\; \min_i {}^i k_{j-1}, \qquad (5.16)$$

the control error $\tilde{z}$ of the "hypothesis" system can be expressed as

$$\begin{aligned} \tilde{z}(t) \;&=\; k_{j-1}\left(x_{j-1,d}(t)-x_{j-1}(t)\right)^{\mathrm{T}} + \dot{x}_{j-1,d}(t) - \dot{x}_{j-1}(t) \\ &=\; k_{j-1}\cdot z(t)^{\mathrm{T}} + \dot{z}(t) \end{aligned} \qquad (5.17)$$

or

$$\dot{z}(t) \;=\; -k_{j-1} \cdot z(t)^{\mathrm{T}} + \tilde{z}(t). \tag{5.18}$$

To show $z(t)$ being GAS, one has to ascertain that for any $\varepsilon > 0$ exists $\tilde{T}$ such that $\|z(t)\| < \varepsilon$ for $t \geq \tilde{T}$. Now, as $\tilde{z}(t)$ is GAS, choose T such that $\|\tilde{z}(t)\| < \frac{1}{2}k\varepsilon/n_{j-1}$ for $t \geq T$.

If for any of $z(t)$'s components $\|^{i}z(t)\| \geq \varepsilon/n_{j-1}$ for some $t \geq T$, it follows[2] for *positive* $^{i}z(t)$

$$
\begin{aligned}
{}^{i}\dot{z}(t) \;&=\; -{}^{i}k_{j-1} \cdot {}^{i}z(t) + {}^{i}\tilde{z}(t) \\
&<\; -k \cdot {}^{i}z(t) + \frac{1}{2}k\varepsilon/n_{j-1} \\
&\leq\; -k\varepsilon/n_{j-1} + \frac{1}{2}k\varepsilon/n_{j-1} \\
&=\; -\frac{1}{2}k\varepsilon/n_{j-1} \\
&<\; 0
\end{aligned}
\tag{5.19}
$$

and for *negative* $^{i}z(t)$

$$
\begin{aligned}
{}^{i}\dot{z}(t) \;&=\; -{}^{i}k_{j-1} \cdot {}^{i}z(t) + {}^{i}\tilde{z}(t) \\
&>\; -k \cdot {}^{i}z(t) - \frac{1}{2}k\varepsilon/n_{j-1} \\
&\geq\; +k\varepsilon/n_{j-1} - \frac{1}{2}k\varepsilon/n_{j-1} \\
&=\; +\frac{1}{2}k\varepsilon/n_{j-1} \\
&>\; 0.
\end{aligned}
\tag{5.20}
$$

Hence, whenever $\|^{i}z(t_0)\| \geq \varepsilon/n_{j-1}$ for some $t_0 \geq T$, it only requires a finite time Δt until $\|^{i}z(t)\| < \varepsilon/n_{j-1}$ for all $t \geq t_0 + \Delta t$. Therefore, this will as well be the case for *all* components i after a finite time, say for $t \geq \tilde{T}$. Then,

$$
\begin{aligned}
\|z(t)\| \;&\leq\; \sum_{i=1}^{n_{j-1}} \|^{i}z(t)\| \\
&<\; n_{j-1} \cdot \varepsilon/n_{j-1} \\
&=\; \varepsilon,
\end{aligned}
\tag{5.21}
$$

which concludes the inductive step for the integrator stage.

Please note that the use of a differentiator element for deriving the change rate input to the "hypothesis" system, as depicted in figure 5.5, is only for clarity reasons. In an actual implementation, the control function of the previous stages can often be differentiated symbolically to calculate $\dot{x}_j$ in closed form.

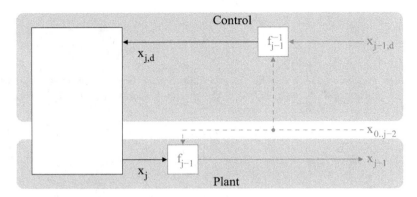

Figure 5.6: Inductive step over a single nonlinear transform stage (f_{j-1}).

- Inductive step over a nonlinear transform stage of figure 5.2: Assuming x_j asymptotically tracks the desired value $x_{j,d}$ (inductive hypothesis), consider an augmented closed-loop system with an additional transform stage f_{j-1} in the "plant" part and an "inverted transform" controller in the "control" part, as depicted by figure 5.6. The control is defined by

$$x_{j,d}(t) = f_{j-1}^{-1}(x_{j-1}(t), x_{j-2}(t), ..., x_0(t), t) \qquad (5.22)$$

with f_{j-1}^{-1} denoting the inverted transform according to (5.9). Defining the control errors as in (5.14,5.15) and using $X(t) = (x_{j-2}(t), ..., x_0(t))$ for compact notation, it is:

$$\tilde{z}(t) = f_{j-1}^{-1}(x_{j-1,d}(t), X(t), t) - x_j(t) \qquad (5.23)$$

$$z(t) = x_{j-1,d}(t) - f_{j-1}(x_j(t), X(t), t)$$

$$= x_{j-1,d}(t) - f_{j-1}\left(f_{j-1}^{-1}(x_{j-1,d}(t), X(t), t) - \tilde{z}(t), X(t), t\right) \quad (5.24)$$

$$\lim_{\tilde{z}(t) \to 0} z(t) = x_{j-1,d}(t) - f_{j-1}\left(f_{j-1}^{-1}(x_{j-1,d}(t), X(t), t), X(t), t\right)$$

$$= x_{j-1,d}(t) - x_{j-1,d}(t)$$

$$= 0 \qquad (5.25)$$

Here, the last step follows from the limit theorems, since f_{j-1}^{-1} is continuous.

This concludes the inductive step for the transform stage. $\square$

The proof as given above emphasizes very clearly how the dynamics aspect (integrator stages) and the nonlinear coupling (transform stages) can – and should – be treated separately to simplify the derivation of a stable control system. The only controller type used above, as far as the dynamics aspect is concerned, is a simple P controller, plus additional open-loop control based on change rate propagation!

---

[2]At this very point of the proof, the use of Lyapunov theory would have helped saving one or two lines of equations.

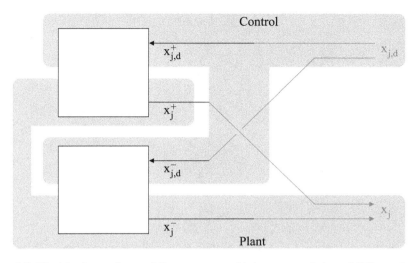

Figure 5.7: "Forking" step for modeling systems with integrator chains of different lengths.

Of course, not all systems to be controlled will show the same number of integrator stages between the control input vector u and all of the target variables in x_0. In order to address this fact, "forks" in the control cascade according to figure 5.7 can be used: Decomposing the actual and desired state vectors at some intermediate stage j as

$$x_j = (x_j^-, x_j^+) \qquad (5.26)$$

and

$$x_{j,d} = (x_{j,d}^-, x_{j,d}^+) \qquad (5.27)$$

serving as inputs and outputs of two separate "upstream" systems, above proof – and all following considerations – can be applied to both of these systems without any change. In particular, the "+" and "−" partial upstream systems may be of different order.

5.3.4   Adaptive Control: Bias Compensation Stage

The proof in the previous section pertains to an *ideal* system the available model of which is *perfect*. While it is not unusual in the literature to examine such cases, any actual implementation along these lines will inevitably fail – this realization is closely connected to the criticism set forth in section 5.2.5. The reason for this failure is that any real control system should be adaptive, at least in the sense of adjusting the controller's operating point to the *actual* one of the system. Just as with classical linear PD in contrast to PID control, this statement is as easy as saying that the asymptotic error cannot be reduced to zero otherwise.

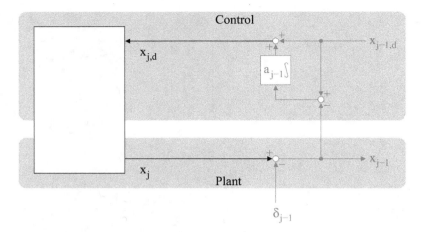

Figure 5.8: Inductive step over a bias compensation stage by control error (unknown but constant bias δ_{j-1}).

Therefore, this section provides an additional inductive step option in the scheme of the proof of proposition 4, introducing a *bias compensation* stage in the "control" part, which compensates for an *unknown but constant* bias in the "plant" part. Bias compensation stages come in two "flavors", the simpler of which is based on tracking error and the more complex one on model error.

Bias Compensation by Tracking Error

This kind of bias compensation stage is depicted in figure 5.8. The constant bias vector is $\delta_{j-1} \in \mathbb{R}^{n_{j-1}}$, and the control is defined by

$$x_{j,d}(t) \;=\; x_{j-1,d}(t) + a_{j-1} \int \left(x_{j-1,d}(t) - x_{j-1}(t) \right)^{\mathrm{T}} dt \tag{5.28}$$

with some gain vector $a_{j-1} \in \mathbb{R}_+^{n_{j-1}}$. Denoting the control errors as defined through (5.14,5.15), the control error of the "hypothesis" system can be expressed as:

$$\begin{aligned}
\tilde{z}(t) &= x_{j-1,d}(t) + a_{j-1} \int \left(x_{j-1,d}(t) - x_{j-1}(t) \right)^{\mathrm{T}} dt - x_{j-1}(t) - \delta_{j-1} \\
&= z(t) + a_{j-1} \int z(t)^{\mathrm{T}} dt - \delta_{j-1}
\end{aligned} \tag{5.29}$$

Now, introducing the derived system

$$\hat{z}(t) \;=\; a_{j-1} \int z(t)^{\mathrm{T}} dt - \delta_{j-1} \tag{5.30}$$

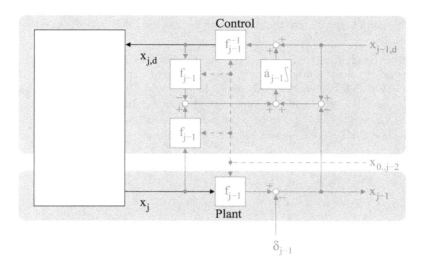

Figure 5.9: Inductive step over a "smart" bias compensation stage by system error (unknown but constant bias δ_{j-1}).

yields

$$\bar{z}(t) \;=\; a_{j-1}^{-1} \cdot \dot{\hat{z}}(t)^{\mathrm{T}} + \hat{z}(t) \tag{5.31}$$

$$\dot{\hat{z}}(t) \;=\; -a_{j-1} \cdot \hat{z}(t)^{\mathrm{T}} + \bar{z}(t) \tag{5.32}$$

with a_{j-1}^{-1} meaning ${}^i a_{j-1}^{-1} = 1/{}^i a_{j-1}$. Here, (5.32) exactly corresponds to (5.18), only for $\hat{z}(t)$ instead of $z(t)$. Hence, applying the same argument as following (5.18) in the inductive step above, one obtains

$$\lim_{t \to \infty} \hat{z}(t) \;=\; 0 \tag{5.33}$$

and via (5.29) finally

$$\lim_{t \to \infty} z(t) \;=\; \lim_{t \to \infty} (\bar{z}(t) - \hat{z}(t)) \;=\; 0, \tag{5.34}$$

which concludes the inductive step for this kind of bias compensation stage. $\square$

Bias Compensation by Model Error

While the bias compensation stage examined above does render the closed-loop system's behavior GAS, its disadvantage is that the corresponding convergence procedure strongly depends on the decay of the control error $\bar{z}(t)$ of the "hypothesis" system, as is clearly expressed by (5.32). Whenever a new desired trajectory $x_{j-1,d}$ is provided, $\bar{z}(t)$ has to converge to zero anew, which in turn may interfere with the convergence of the integrator in the bias compensation stage.

This motivates the application of a "smart" bias compensation stage more directly guided by the model error. The model error, expressed by δ_{j-1}, cannot usually be measured directly due to a lack of observability. Therefore, the "smart" bias compensation is specialized to include both a bias stage $\delta_{j-1} \in \mathbb{R}^{n_{j-1}}$ and a prior transform stage f_{j-1} in the "plant" branch, as shown in figure 5.9. It is assumed that x_j can be measured and fed forward through an f_{j-1} model in the "control" branch. Using the naming convention from the previous subsections, the control is defined by

$$x_{j,d}(t) \;=\; f_{j-1}^{-1}\left(x_{j-1,d}(t) + \gamma(t), X(t), t\right) \tag{5.35}$$

$$
\begin{aligned}
\gamma(t) \;=\;& a_{j-1} \int \left(x_{j-1,d}(t) - f_{j-1}\left(x_{j,d}(t), X(t), t\right)\right. \\
& \left. - x_{j-1}(t) + f_{j-1}\left(x_j(t), X(t), t\right)\right)^{\mathrm{T}} dt \\
\;=\;& a_{j-1} \int \left(-\gamma(t) - x_{j-1}(t) + x_{j-1}(t) + \delta_{j-1}\right)^{\mathrm{T}} dt \\
\;=\;& a_{j-1} \int \left(\delta_{j-1} - \gamma(t)\right)^{\mathrm{T}} dt, \tag{5.36}
\end{aligned}
$$

which can be used to express a derived system $\hat{z}(t)$ analog to (5.30):

$$\hat{z}(t) \;=\; \gamma(t) - \delta_{j-1} \tag{5.37}$$

$$
\begin{aligned}
\dot{\hat{z}}(t) \;=\;& a_{j-1}\left(\delta_{j-1} - \gamma(t)\right)^{\mathrm{T}} \\
\;=\;& -a_{j-1}\hat{z}^{\mathrm{T}}(t) \tag{5.38}
\end{aligned}
$$

Hence, with $\bar{z}(t)$ denoting the control error of the subsystem up to the transform stage, it is

$$\bar{z}(t) + \hat{z}(t) \;=\; z(t) \;=\; x_{j-1,d}(t) - x_{j-1}(t), \tag{5.39}$$

yielding

$$\lim_{t \to \infty} z(t) \;=\; 0 \tag{5.40}$$

because of proposition 4 (transform stage) and $\hat{z}(t) \to 0$ due to (5.38). $\square$

Most notably, the convergence of $\hat{z}(t)$ does *not* depend on $\bar{z}$ in this case, but constitutes a direct exponential decay. This will result in better quantitative performance of the closed-loop system whenever $x_{j,d}$ is not directly (i.e. functionally) coupled to the plant's control input u.

Looking at figure 5.9, one might argue that the "smart" bias compensation could be equally implemented in a much simpler way: With ideally inverted transform functions, the $x_{j,d}$ and $x_{j-1,d}$ inputs to the integrator just provide a negative feedback equal to $-\gamma(t)$, while the x_j and x_{j-1} inputs are together equal to $+\delta_{j-1}$. Thus, $\gamma(t)$ is nothing but a first-order low-pass filter on δ_{j-1}. This is of course exactly the statement of (5.38). Therefore, $\gamma(t)$ could as well be fed back directly, canceling the necessity of using the two integrator inputs from the "desired" branch.

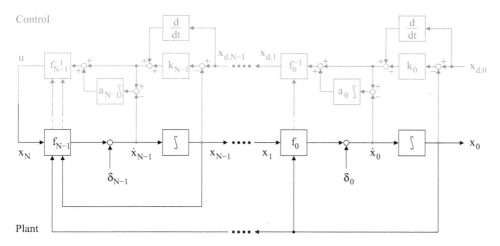

Figure 5.10: Closed-loop adaptive control system, resulting from proposition 4 and bias compensation.

Yet, in any actual implementation, the transform stage models f_{j-1} and f_{j-1}^{-1} used by the controller will *not* be ideal. Then, the asymptotic error of $z(t)$ in the "optimized" implementation of figure 5.9 will not in general vanish as desired. On the other hand, the depicted implementation guarantees that the integrator input $\dot{\gamma}(t)$ is exactly equal to the simpler bias compensation stage according to figure 5.8 *whenever* $x_{j,d} = x_j$. For then, two identical signals are fed through the same transform model f_{j-1} and subtracted afterwards, yielding zero integrator input "from the left-hand side". Hence, the behavior of a bias-compensated control system implemented according to figure 5.9 will be truly GAS even in the presence of rounding errors and non-optimally modeled transform functions.

Closed-Loop View

Bias compensation stages according to figure 5.8 can be freely introduced into any system model according to figure 5.2. Considering the fact that the transform stages according to figure 5.6 do *not* use the local control error $z(t)$ in their "control" part, the adding of these compensation stages is most reasonable done, with respect to observability, on the "right-hand side" (regarding figure 5.2) of every transform stage. The resulting class of closed-loop control systems is summarized by figure 5.10. This illustration shows how the "control" part iteratively computes the desired state variables $x_{i,d}$ based on the observed system state and response. It uses "simple" bias compensation stages only, for clarity reasons.

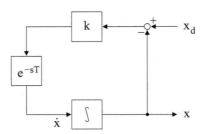

Figure 5.11: Minimal closed-loop system for investigating the effect of a dead time T.

5.3.5 Destabilizing Effect of Dead Time

The stability characteristics proven in the previous sections 5.3.3 and 5.3.4 do not, in any way, depend on the choice of gain vectors (k_j for the integrator stage and a_j for bias compensation). The class of systems examined there does not, however, incorporate potential dead time in its dynamics. Of course, the combination of dead time and integrators in the closed loop may lead to instability, depending on the gain vectors used. While it is out of the scope of this book to formally investigate the effect of dead time within a full nonlinear cascade control system like 5.7, this section derives an upper bound on the gain subject to the dead time within the closed loop, using a maximally simplified linear system as a substitute.

Figure 5.11 depicts this substitute system. It is a closed-loop system with scalar input and output, consisting of a gain $k \in \mathbb{R}$, a dead time $T > 0$, and an integrator. Here, the gain k may represent some control gain $^i k_j$ from an integrator stage (figure 5.5) as well as some bias compensation gain $^i a_j$ from a bias compensation stage (figure 5.8). This system will be analyzed through Laplace transform and transfer functions, i.e. the "traditional" tools for linear systems.

The resulting closed-loop transfer function between $G(s) = X(s)/X_d(s)$ is:

$$G(s) = \frac{\frac{k}{s} e^{-sT}}{1 + \frac{k}{s} e^{-sT}} \tag{5.41}$$

Therefore, the system is stable if and only if the parameters k and T are set such that all solutions of

$$s + k e^{-sT} = 0 \tag{5.42}$$

have $\text{Re}(s) < 0$ (transformation of the denominator possible since $s = 0$ is no pole of G). The solution of (5.42) is not fully analytic, but can be obtained for this special case with the substitution $s = x + yi$ and writing down the real and imaginary parts separately:

$$x + k e^{-xT} \cos -yT = 0 \tag{5.43}$$
$$y + k e^{-xT} \sin -yT = 0 \tag{5.44}$$

For solutions with $x \geq 0$, (5.43) clearly requires

$$\|yT\| \geq \frac{\pi}{2} \qquad \Rightarrow \qquad \|y\| \geq \frac{\pi}{2T}, \tag{5.45}$$

and then (5.44) yields, because of $\|e^{-xT} \sin -yT\| \leq 1$:

$$k \; \geq \; \|y\| \; \geq \; \frac{\pi}{2T} \tag{5.46}$$

Therefore, no solutions in the right half-plane (including $x = 0$) exist for

$$k < \frac{\pi}{2T} \qquad \Leftrightarrow \qquad T < \frac{\pi}{2k} \tag{5.47}$$

On the other hand, if $T = \frac{\pi}{2k}$, there are two solutions $s_1 = -ki$, $s_2 = +ki$ to (5.42) with non-negative real part.

(5.47) provides a simple and convenient way of estimating an absolute upper bound on the gains, especially those for bias compensation. Nevertheless, it has to be kept in mind that this bound has been derived from a massively simplified model. The higher the actual order of the system between x_d and x as used in this examination is, the more safety margin may be required in selecting the gains.

5.3.6 Resorting to Input-Output Stability

So far, the stability results obtained have referred to global asymptotic stability, or zero tracking error. This strong class of results was mainly due to the assumption of *constant* model biases in section 5.3.4. In this section, time-varying disturbances are permitted and the proof from sections 5.3.3 and 5.3.4 repeated. However, the achievable stability result will be weakened to global input-output stability.

Proposition 5 *Given a closed-loop control system according to figure 5.10, permitting time-varying, differentiable biases $\delta_j(t)$ with $\dot{\delta}_j(t)$ continuous and bounded, and a desired trajectory $x_{0,d}(t)$ continuously differentiable at least N times, all of its resulting solutions*

$$z(t) \; = \; x_0(t) - x_{0,d}(t) \tag{5.48}$$

are GIOS.

Proof of proposition 5: This proof is structured corresponding to the proof of proposition 4 and re-uses expressions from the latter wherever possible. The inductive basis (5.12) can be adopted without change, since any GAS solution is GIOS as well. The inductive step again distinguishes between the different construction stages:

- Inductive step over an integrator stage (figure 5.5): One has to ascertain that there exists $\varepsilon^* \in \mathbb{R}_+$ such that $\|z(t)\| < \varepsilon^*$ for $t \geq t_0$. With $\tilde{z}$ from (5.18) being GIOS (inductive hypothesis), ε can be set sufficiently large to again fulfill $\|\tilde{z}(t)\| < \frac{1}{2}k\varepsilon/n_{j-1}$, but for $t \geq t_0$. With the same argument as subsequent to (5.18) above, it follows that $\|z(t)\| \leq \varepsilon$ after a finite time, say for $t \geq \tilde{T}$. The actual bound ε^* can then safely be taken as:

$$\varepsilon^* \quad = \quad 2 \cdot \max \left\{ \varepsilon \right\} \cup \left\{ \|z(t)\| \mid t_0 \leq t < \tilde{T} \right\} \tag{5.49}$$

  Here, the maximum exists because none of the components $^i z(t)$ can have escaped to infinity.

- Inductive step over a transform stage (figure 5.6): Looking at (5.24), the boundedness of $z(t)$ follows from the boundedness of $\tilde{z}(t)$ (inductive hypothesis) and the boundedness of the Jacobian of f_{j-1}.

- Inductive step over a bias compensation stage (figure 5.8 but with time-varying $\delta_{j-1}(t)$): Due to the non-zero derivative of $\delta_{j-1}(t)$, (5.32) transforms into

$$\dot{\hat{z}}(t) \quad = \quad -a_{j-1} \cdot \hat{z}(t)^{\mathrm{T}} + \tilde{z}(t) - \dot{\delta}_{j-1}(t), \tag{5.50}$$

  thus $\hat{z}(t)$ is GIOS since (5.50) corresponds to (5.18), both $\tilde{z}(t)$ (inductive hypothesis) and $\dot{\delta}_{j-1}(t)$ (by assumption) are bounded, and hence, the argument used above for the integrator stage can be applied here as well. $\square$

Proposition 5 is particularly important in assuring that certain remaining model errors and dead times in the system under investigation cannot critically affect its practical stability characteristics. Instead, proposition 5 indicates that system behavior will only gradually degrade from zero tracking error to bounded tracking error.

This concludes the formal investigation of the class of control systems employed in this book. The next section examines a special subclass thereof and provides compact, closed-form solutions for control laws and tracking trajectories.

Up to now, the boundedness of control inputs and parts of the state variables, or in other words the possibility of actuator and rate saturation, which is present in all practical systems, has not been considered. Section 5.4.5 below will introduce a *mode switching* approach that is sufficient to arrange for actual global stability of $x_{0,d}(t) - x_0(t)$ in practical control systems.

5.4   Multi-Integrator Cascades

An important special case of the control systems examined in the previous section, which arises naturally and permits significant simplification, consists of cascades of just N integrators, $N \geq 1$, without transform and bias stages in between, or equivalently, *identity* transforms only. This kind of control systems is examined in some detail in this section. For compactness of notation, only scalar state variables $x_j(t) \in \mathbb{R}$ will be considered here. However, all formulas presented in this section immediately generalize to state vectors as well.

An integrator cascade of at least second order is encountered in almost all cases of UAV flight control, because these systems usually incorporate (at least) the two integration stages from acceleration to position. In this part of the system, it is also always true that there is no additional transform involved between the two integrator stages.

In particular, this section presents three different views on this class and proves their equivalence. These views are

1. a differential equation in the tracking error signal,

2. the closed-form representation of the control law, and

3. the closed-form representation of the convergence trajectory in the time domain,

each of which is dedicated one of the following three subsections.

The basis of all these views is the integrator stage as introduced in figure 5.5, with its control part defined, basically repeating (5.13), by:

$$x_{j+1,d}(t) \quad = \quad k_j \left(x_{j,d}(t) - x_j(t) \right) + \dot{x}_{j,d}(t) \tag{5.51}$$

Furthermore, it is $x_i = x_0^{(i)}$ for $0 \leq i \leq N$.

After the discussion of the three views, two more subsections deal with the choice of suitable gains k_j (section 5.4.4) and with multi-mode control when using distant control targets (section 5.4.5).

While all the results presented in this section are basically known facts from the theory of linear control systems, they are not often written down explicitly for an arbitrary system order N, nor for trajectories in the time domain. This should suffice as a motivation for this section.

5.4.1   Differential Equation View

(5.51) can be transformed into:

$$\begin{aligned} x_{j+1,d}(t) - x_{j+1}(t) \quad &= \quad k_j \left(x_{j,d}(t) - x_j(t) \right) + \dot{x}_{j,d}(t) - \dot{x}_j(t) \\ &= \quad k_j \left(x_{j,d}(t) - x_j(t) \right) + \frac{d}{dt} \left(x_{j,d}(t) - x_j(t) \right). \end{aligned} \tag{5.52}$$

Using operator notation $Dx(t) := \frac{d}{dt}x(t)$ and thanks to the linearity of differentiation, this can be rewritten as:

$$x_{j+1,d}(t) - x_{j+1}(t) \quad = \quad (D + k_j) \left(x_{j,d}(t) - x_j(t) \right) \tag{5.53}$$

Now, applying (5.53) repeatedly for all stages, one obtains a differential equation in "polynomial" notation, i.e.:

$$x_{N,d}(t) - x_N(t) \quad = \quad \left(\prod_{i=0}^{N-1} (D + k_i) \right) \left(x_{0,d}(t) - x_0(t) \right) \tag{5.54}$$

When considering to apply (5.54) practically, please keep in mind that $-x_N$ appears on *both* sides of this equation, so it can (and must) be omitted when actually computing $x_{N,d}(t)$.

In the special case that $x_{N,d}$ equals the control input $u = x_N$, i.e. if the integrator cascade resides directly at the input stage of the system, (5.54) can be written more compact as:

$$\left(\prod_{i=0}^{N-1} (D + k_i) \right) \left(x_{0,d}(t) - x_0(t) \right) = 0 \tag{5.55}$$

5.4.2   Control Law View

Proposition 6 *The closed-form explicit control law for an Nth-order integrator cascade according to figure 5.5 and equation (5.51) is given by:*

$$x_{N,d}(t) = x_{0,d}^{(N)}(t) + \sum_{i=0}^{N-1} \left(\sum_{\substack{S \subseteq \{0;\dots;N-1\} \\ |S| = N-i}} \prod_{j \in S} k_j \right) \left(x_{0,d}^{(i)}(t) - x_i(t) \right) \tag{5.56}$$

In order to make intuitively clear what (5.56) expresses, consider the case $N = 3$ as an example. Abbreviating $x = x_0$, $x_d = x_{0,d}$, and $z = x_0 - x_{0,d}$, it is

$$
\begin{aligned}
x_{3,d} = \; & x_d^{(3)} \\
+ \; & (k_0 + k_1 + k_2)\, \ddot{z} \\
+ \; & (k_0 k_1 + k_1 k_2 + k_0 k_2)\, \dot{z} \\
+ \; & k_0 k_1 k_2\, z
\end{aligned}
\tag{5.57}
$$

Proof of proposition 6: Looking at (5.54), the operators $(D + k_i)$ can be multiplied out just like an ordinary product, due to the linearity of differentiation and scaling. Thus, the right side of (5.54) consists of the following summands:

1. $x_{0,d}^{(N)} - x_N$, through the D^N-part of the operator term,

2. for each $1 \leq i \leq N$, all summands consisting of some combination of i of the k-factors, applied to the $(N-i)$th derivative of $x_{0,d} - x_0$.

After removing x_N from both sides of the equation, this exactly equals (5.56). $\square$

One crucial property of (5.56) is that it is fully symmetric in the ordering of the k_j. This means that it is of no consequence at all to the resulting control system how a set of gains is distributed over the sequence of integrator stages, while the intermediate "desired" values $x_{j,d}(t)$, $0 < j < N$, still *do* depend on the ordering of the gains nevertheless. Furthermore, (5.57) indicates that the resulting controllers have two, and not one, degrees of freedom, namely receiving $x_d(t)$ and $z(t) = x_d(t) - x(t)$ as two different input signals.

In the special case that $x_{N,d}$ equals the control input $u = x_N$, (5.56) can be written more compact as

$$\sum_{i=0}^{N} \left(\sum_{\substack{S \subseteq \{0;\ldots;N-1\} \\ |S| = N-i}} \prod_{j \in S} k_j \right) \left(x_{0,d}^{(i)}(t) - x_i(t) \right) = 0, \tag{5.58}$$

by including $x_{0,d}^{(N)}(t) - x_N(t)$ in the outer sum and making use of the convention that an empty product equals 1.

5.4.3 Trajectory View

Proposition 7 *A system consisting of an Nth-order integrator cascade and associated controllers according to figure 5.5 and equation (5.51) and its control input $u = x_{N,d} = x_N$ directly provided by the series of controllers behaves in the time domain according to*

$$x_{j,d}(t) - x_j(t) = \sum_{k \in \{k_0;\ldots;k_{N-1}\}} P_k^j(t)\, e^{-kt}, \tag{5.59}$$

for all $0 \le j \le N$, with P_k^j being polynomials in $\mathbb{R}$ and

$$\deg P_k^j = \|\{i \ge j \mid k_i = k\}\| - 1, \tag{5.60}$$

assuming $\deg P = -1$ iff $P(t) = 0$.

This proposition does *not* explicitly state the resulting weighting polynomials. Instead, it expresses only the structure of the resulting trajectory equations in the time domain.

In order to make the statement of proposition 7 more intuitively clear, consider the case $N = 3$, $j = 0$ as an example. As the degree of the weighting polynomials P_j^k depends on the number of equal k-factors, there are three different cases to distinguish:

$$k_0 \ne k_1 \ne k_2 \ne k_0: \quad x_d(t) - x(t) = Ae^{-k_0 t} + Be^{-k_1 t} + Ce^{-k_2 t} \tag{5.61}$$
$$k_0 = k_1 \ne k_2: \quad x_d(t) - x(t) = (D + Et)\, e^{-k_0 t} + Fe^{-k_2 t} \tag{5.62}$$
$$k_0 = k_1 = k_2: \quad x_d(t) - x(t) = (G + Ht + It^2)\, e^{-k_0 t} \tag{5.63}$$

Here, each triple of constants out of $A, B, C, D, E, F, G, H, I \in \mathbb{R}$ depends on the boundary values $x(t_0)$, $\dot{x}(t_0)$, and $\ddot{x}(t_0)$.

Proof of proposition 7: By induction over j, from $j = N$ to $j = 0$.

Inductive basis $j = N$:

$$x_{N,d}(t) - x_N(t) = 0 \tag{5.64}$$

follows from $x_N = x_{N,d}$, by assumption, and equals (5.59) because of $\forall k : P_k^N(t) = 0$.
Inductive step $j+1 \rightarrow j$: Using (5.51) and inserting the inductive hypothesis yields

$$
\begin{aligned}
k_j\,(x_{j,d}-x_j(t)) &= x_{j+1,d}(t)-\dot{x}_{j,d}(t) \\
&= x_{j+1}(t)-\dot{x}_{j,d}(t)+x_{j+1,d}(t)-x_{j+1}(t) \\
&= x_{j+1}(t)-\dot{x}_{j,d}(t)+ \sum_{k\in\{k_0;\ldots;k_{N-1}\}} P_k^{j+1}(t)\,e^{-kt} \qquad (5.65)
\end{aligned}
$$

Substituting $z_j = x_{j,d}-x_j$ for the jth stage's tracking error, this becomes:

$$
k_j\cdot z_j(t) = -\dot{z}_j(t)+ \sum_{k\in\{k_0;\ldots;k_{N-1}\}} P_k^{j+1}(t)\,e^{-kt} \qquad (5.66)
$$

$$
z_j(t) = -\frac{1}{k_j}\left(\dot{z}_j(t)-\sum_{k\in\{k_0;\ldots;k_{N-1}\}} P_k^{j+1}(t)\,e^{-kt}\right) \qquad (5.67)
$$

This is a differential equation in $z_j(t)$. It can be solved by a function according to the stage-j equation from (5.59) with properly determined P_k^j as follows. Grouping (5.67) according to the different e^{-kt}-terms yields the following set of equations (for $k \in \{k_0;\ldots;k_{N-1}\}$):

$$
P_k^j(t) = -\frac{1}{k_j}\left(\dot{P}_k^j(t)-kP_k^j(t)-P_k^{j+1}(t)\right) \qquad (5.68)
$$

Now, there are two cases to distinguish: For $k = k_j$, the P_k^j terms drop out of above equation, leaving just:

$$
\dot{P}_k^j(t) = P_k^{j+1}(t) \qquad (5.69)
$$
$$
P_k^j(t) = \int P_k^{j+1}(t)dt + C^j \qquad (5.70)
$$

Here, the constant of integration C^j can be chosen to fulfill some initial condition $z_j(t_0)$, and P_k^j is a polynomial with a degree one higher than P_k^{j+1}, as permitted by (5.60) in this case. If, as a special case, k_j is the "first" occurrence of k as an exponent, P_k^j will only consist of the constant of integration, shifting the polynomial's degree from -1 to 0.
For $k \neq k_j$, however, (5.68) is a full differential equation:

$$
(k_j-k)P_k^j(t) = P_k^{j+1}(t)-\dot{P}_k^j(t) \qquad (5.71)
$$
$$
(k_j-k)P_k^j(t)+\dot{P}_k^j(t) = P_k^{j+1}(t) \qquad (5.72)
$$

Now writing

$$
n = \deg P_k^{j+1} \qquad (5.73)
$$
$$
P_k^j(t) = \sum_{i=0}^n c_i^j\cdot t^i \qquad (5.74)
$$
$$
P_k^{j+1}(t) = \sum_{i=0}^n c_i^{j+1}\cdot t^i, \qquad (5.75)
$$

(5.72) is solved by

$$(k_j - k)c_n^j = c_n^{j+1} \tag{5.76}$$

$$c_n^j = \frac{1}{k_j - k} c_n^{j+1} \tag{5.77}$$

and subsequent iteration over

$$(k_j - k)c_i^j + (i+1)c_{i+1}^j = c_i^{j+1} \tag{5.78}$$

$$c_i^j = \frac{1}{k_j - k}\left(c_i^{j+1} - (i+1)c_{i+1}^j\right) \tag{5.79}$$

for $n > i \geq 0$, which uniquely determines all coefficients. Here, P_k^j and P_k^{j+1} are of equal degree, as required by (5.60). This concludes the $k \neq k_j$ case. $\square$

5.4.4 Choice of the k-Gains

Apart from the interaction with dead time in the system as discussed in section 5.3.5, the qualitative convergence properties of any system according to figure 5.10 do in no way depend on the choice of the gains k_j. However, there will be bounds on the control inputs in any practical system. In order to meet these in any desirable state $x(t)$, a simple examination of the resulting control laws in the form of (5.56) suffices to choose the k_j low enough – not uniquely, though.

This section provides a more detailed analysis of the effects of choosing the gains. This is done separately for the smallest k_j in the control cascade on the one hand, and for the remaining ones on the other.

Long-Term Decay of Tracking Error

Looking at the trajectory equation (5.59), it is immediately clear that the control error $x_{0,d}(t) - x_0(t)$ will finally be dominated by some term

$$d(t) = A \cdot t^m e^{-k_{min}t} \tag{5.80}$$

with k_{min} denoting the smallest gain,

$$k_{min} = \min_{j \in \{0,\dots,N-1\}} k_j, \tag{5.81}$$

and m being the largest exponent in the polynomial $P_{k_{min}}^0$. As the exponential decay in (5.80), in turn, will be faster in the limit than the increase of the power of t, the half-life period $T_{1/2}$ of this dominating decay suitably characterizes the system's long-term tracking convergence, with

$$T_{1/2} = \frac{\ln 2}{k_{min}}. \tag{5.82}$$

Therefore, (5.82) provides a good guideline for choosing an *absolute lower bound* on the gains k_j.

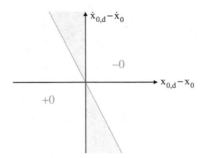

Figure 5.12: Critical plane describing the direction of final convergence, indicated through $+0$ and -0, for $N = 2$.

Critical Hyperplane in State Space

Now, it remains to be seen how the remaining gains contribute to the convergence properties. When looking at the exponential decay characteristics of the trajectory equation (5.59) and the dominating term (5.80), it turns out that the resulting practical convergence trajectories are *usually* free from overshooting. Informally speaking, overshooting only occurs when the state vector combines a "low" initial tracking error $x_{0,d}(t) - x_0(t)$ and "high" initial values of the derivatives $x_{0,d}^j(t) - x_j(t)$ for $j \geq 1$. In practice, this case can (and should) be usually avoided.

With respect to overshooting, the *direction* of the final approach to the target $x_{0,d}(t)$ is of particular relevance. This direction is practically reflected in the sign of A in the term (5.80) above. The following proposition formally introduces a *critical hyperplane* that specifies the final approach direction, using all k_j-gains except (one instance of) $k_{\min}$.

Proposition 8 *Consider an Nth-order integrator cascade in closed-loop with a controller according to (5.56) and the gains k_j sorted such that*

$$k_{\min} = k_{N-1} \ \leq \ k_{N-2} \leq \ ... \ \leq \ k_0. \tag{5.83}$$

Then, the (final) convergence direction of the tracking error obeys:

$$x_{0,d}(t) - x_0(t) \ \rightarrow \ -0 \quad \text{if} \quad x_{N-1,d}(t_0) - x_{N-1}(t_0) > 0 \tag{5.84}$$
$$x_{0,d}(t) - x_0(t) \ \rightarrow \ +0 \quad \text{if} \quad x_{N-1,d}(t_0) - x_{N-1}(t_0) < 0 \tag{5.85}$$

Please note that the sorting of the gains does not affect the controller here, because (5.56) is symmetric in the gains. The term $x_{N-1,d}(t_0)$ above is to be calculated according to the control law (5.56) for one integrator stage before the controller's output $x_{N,d} = u$. For the actual value of $x_{N-1,d}(t_0)$, it *is* relevant indeed that $k_{N-1} = k_{\min}$ is excluded from the sequence of gains.

The inequalities in (5.84) and (5.85) define an hyperplane in the N-dimensional space of state errors $D^j(x_{j,d} - x_j)$ with $0 \leq j \leq N - 1$, separating initial conditions resulting in "positive" and "negative" convergence. Consider the case $N = 2$ as an example to make this fact more intuitively clear. In this case, the condition from (5.85) reads

$$
\begin{aligned}
0 &< k_0 \cdot (x_{0,d} - x_0) + \dot{x}_{0,d} - x_1 \\
&= k_0 \cdot (x_{0,d} - x_0) + D(x_{0,d} - x_0).
\end{aligned}
\tag{5.86}
$$

Figure 5.12 depicts this critical hyperplane in the space of state errors. The shaded areas indicate the undesirable zone in which the sign of the initial x_0 error differs from the sign of the final approach, i.e. where overshooting occurs during convergence. The slope of the hyperplane is given by k_0 in this case, which in turn is the maximum of $\{k_0, k_1\}$. The higher k_0 is chosen, the smaller are the undesirable areas with overshooting.

Assuming constant change rates in (5.86), i.e. zero control input $u = 0$, the critical hyperplane can also be regarded as the set of initial states that would lead to $x_0 = x_{0,d}$ after exactly $T = 1/k_0$ time with the controller "inactive".

Please note that for $N > 2$, proposition 8 does not formally guarantee the absence of overshooting because the trajectory may have multiple zero points before convergence then. However, in practice this would only actually happen with fairly extreme combinations of the initial error derivatives. Conditions of this kind can usually be safely excluded from the state space the controller operates in.

Proof of proposition 8: It suffices to examine the sign of A in (5.80): $A < 0$ corresponds to (5.84), while $A > 0$ corresponds to (5.85).

Let

$$
P_*^j(t) = A_*^j t^m
\tag{5.87}
$$

denote the lowest k-gain, highest t-power summand, analog to (5.80), in some stage's trajectory equation (5.59). In the following, it is shown via induction from $j = N - 1$ to $j = 0$ that

$$
\operatorname{sgn} A_*^j = \operatorname{sgn} \left(x_{N-1,d}(t_0) - x_{N-1}(t_0) \right).
\tag{5.88}
$$

Inductive basis $j = N - 1$: According to (5.59), it is

$$
x_{N-1,d}(t) - x_{N-1}(t) = A_*^{N-1} \cdot e^{-k_{\min} t},
\tag{5.89}
$$

hence

$$
\operatorname{sgn} A_*^{N-1} = \operatorname{sgn} \left(x_{N-1,d}(t_0) - x_{N-1}(t_0) \right).
\tag{5.90}
$$

Inductive step $j + 1 \to j$: According to (5.54), it is

$$
\begin{aligned}
P_{k_{\min}}^{j+1} &= (D + k_j) \, P_{k_{\min}}^j(t) \\
&= \left(D P_{k_{\min}}^j - k_{\min} P_{k_{\min}}^j + k_j P_{k_{\min}}^j \right) \\
&= \left(D P_{k_{\min}}^j + (k_j - k_{\min}) P_{k_{\min}}^j \right).
\end{aligned}
\tag{5.91}
$$

Now, there are two cases to distinguish:

1. $k_j = k_{\min}$: From

$$P_{k_{\min}}^{j+1} = DP_{k_{\min}}^j \qquad (5.92)$$

it follows for the highest-degree term that

$$P_*^j(t) = A_*^{j+1}/(m+1)\, t^{m+1}. \qquad (5.93)$$

2. $k_j > k_{\min}$: From (5.91) it follows that P_*^j must be of the same degree as P_*^{j+1}, thus

$$P_*^j(t) = A_*^{j+1}/(k_j - k_{\min})\, t^m. \qquad (5.94)$$

In both cases, it is sgn $A_*^j =$ sgn A_*^{j+1}. The case $k_j < k_{\min}$ cannot occur due to the sorting condition (5.83) above. □

Proposition 8 does not cover the convergence behavior for initial states *exactly* in the critical hyperplane. The structure of the proof, however, implies how these cases depend on hyperplanes in smaller subspaces (without formal analysis in this book).

5.4.5 Distant Target Points

The integrator cascade controller according to (5.56) can very well be used with a considerable initial tracking error, that is, with occasional jumps in the desired value $x_{0,d}$ instead of utilizing a smooth desired trajectory that permanently minimizes the tracking error. This issue is to be discussed in the current section.

On the one hand, using a smooth desired trajectory will result in "zero" tracking error permanently, according to proposition 4. In this case, the actual trajectory followed by the system can be deterministically chosen by the user. On the other hand, it is very challenging to expect some "user" to define a desired trajectory $x_{0,d}(t)$ that is both suitable for the particular application and often-enough differentiable. If this "user" is actually a human, he or she will most probably prefer to restrict the trajectory specification to a set of way-points, largely. For a full general trajectory specification $x_{0,d}(t)$ is just much too much information to be suitably provided through any user interface. Furthermore, the most "natural" trajectories, like combinations of circular arcs and straight lines or ramp functions for acceleration and deceleration, do not fulfill the differentiability requirement. Therefore, the generation of the desired trajectory should honestly be regarded as a part of the controller, instead of as the job of the "user". The latter view actually evades a substantial part of the problem by declaring it a part of the interface.

Hence, this book promotes the use of single way-points (target points $x_{0,d}$) as the primary input to the controller. Proposition 4 guarantees the final convergence to any target point, while the behavior after a jump in $x_{0,d}(t)$ is not as deterministic as with smooth desired trajectories. But the critical hyperplane criterion introduced in section 5.4.4 even precludes overshooting whenever the system is sufficiently stationary at the moment of changing the current target point.

However, a major issue in using distant targets is that the control inputs to the system, as reflected e.g. through the control law (5.56), will grow unbounded with the initial target distance $x_{0,d}(t_0) - x_0(t_0)$. This is a consequence of the first controller stage's demanding a desired velocity $x_{1,d}$ proportional to the target distance. Hence, this approach is *not* directly applicable in any practical system. Nevertheless, it suffices to add a simple rule of *mode switching* to overcome this conflict, as detailed below.

Mode Switching Control

The solution to the bounds problem stated above consists in discriminating two phases (*modes*) of control:

1. A *traveling mode*, applied when the current target distance $x_{0,d}(t) - x_0(t)$ is high, in which "velocity" control replaces "position" control. The desired traveling velocity $\bar{x}_{1,d} > 0$ is constant and constitutes a parameter of the mode-switching controller.

2. An *approach mode*, applied when the current target distance $x_{0,d}(t) - x_0(t)$ is low, in which standard "position" control according to (5.56) occurs.

Using the operator notation as in (5.54) for maximum compactness of notation, the control for approach mode is given by

$$x_{N,d} = x_N + \left(\prod_{i=0}^{N-1} (D + k_i) \right) (x_{0,d} - x_0), \tag{5.95}$$

while the control for traveling mode shall be given by

$$x_{N,d}^{+} = x_N + \left(\prod_{i=1}^{N-1} (D + k_i) \right) (+\bar{x}_{1,d} - x_1) \tag{5.96}$$

$$x_{N,d}^{-} = x_N + \left(\prod_{i=1}^{N-1} (D + k_i) \right) (-\bar{x}_{1,d} - x_1) \tag{5.97}$$

for traveling in positive and negative direction, respectively. It can be easily seen that (5.96, 5.97) are standard controllers for an integrator cascade of order $N - 1$, with the indices renamed to refer to the corresponding state variables and gains from (5.95).

In order to issue a smooth control input $u(t)$ even in the event of a mode change, the proposed formalization of the switching process is the following:

$$u(t) = \begin{cases} x_{N,d}^{-}(t) & \text{if } x_{N,d}(t) < x_{N,d}^{-}(t) \\ x_{N,d}(t) & \text{if } x_{N,d}^{-}(t) \leq x_{N,d}(t) \leq x_{N,d}^{+}(t) \\ x_{N,d}^{+}(t) & \text{if } x_{N,d}(t) > x_{N,d}^{+}(t) \end{cases} \tag{5.98}$$

In this way, exceedingly large position control signals select either positive or negative traveling control instead, depending on the position control signal's sign. However, it is not formally obvious that (5.98) cannot cause oscillations in the selected mode, which might render the closed-loop system unstable. This is to be answered in the next subsection.

Stability

The stability properties of mode-switching control for distant target points are addressed by the following proposition.

Proposition 9 *The solutions of an Nth-order integrator cascade system in closed loop with a mode-switching controller according to (5.98) and directly issuing the control input $u = x_N = x_{N,d}$ are GAS.*

Proof of proposition 9: In order to simplify the proof, it is assumed that $x_{0,d} = 0$. This can easily be relaxed to any constant $x_{0,d}(t)$ or even any constant $\dot{x}_{0,d}(t)$ through a coordinate shift without affecting the result.

First, the sets of states at the boundaries of the mode conditions in (5.98) are described by:

$$
\begin{aligned}
0 &= x_{N,d} - x_{N,d}^{\pm} \\
&= \left(\prod_{i=0}^{N-1} (D+k_i) \right) (x_{0,d} - x_0) - \left(\prod_{i=1}^{N-1} (D+k_i) \right) (\pm \bar{x}_{1,d} - x_1) \\
&= \left(\prod_{i=1}^{N-1} (D+k_i) \right) x_1 - \left(\prod_{i=0}^{N-1} (D+k_i) \right) x_0 - \left(\prod_{i=1}^{N-1} k_i \right) (\pm \bar{x}_{1,d}) \qquad (5.99)
\end{aligned}
$$

$$
\begin{aligned}
\left(\prod_{i=1}^{N-1} k_i \right) (\pm \bar{x}_{1,d}) &= \left(\prod_{i=1}^{N-1} (D+k_i) \right) (x_1 - (D+k_0)x_0) \\
&= -k_0 \left(\prod_{i=1}^{N-1} (D+k_i) \right) x_0 \qquad (5.100)
\end{aligned}
$$

(5.100) describes two parallel hyperplanes in the system's state space $\mathbb{R}^N$, one is the boundary surface of positive traveling and one is the boundary surface of negative traveling. Positive traveling is selected when the RHS of (5.100) is greater than the positive LHS, negative traveling occurs when the RHS is lower than the negative LHS.

With $X = (x_0, ..., x_{N-1})^{\mathrm{T}} \in \mathbb{R}^N$ expressing a state, (5.100) can be represented as

$$
\pm d = E \cdot X \qquad (5.101)
$$

with constant d and a constant normal vector E collecting the RHS coefficients. The system's tendency to cross one of the boundary hyperplanes can now be investigated on behalf of the scalar product $E \cdot \dot{X}$, because

$$
\frac{d}{dt}(E \cdot X) = E \cdot \dot{X} \qquad (5.102)
$$

In the case of approach mode, because of $u = x_{N,d}$ by assumption, it is

$$\dot{X} = \begin{pmatrix} x_1 \\ x_2 \\ \vdots \\ x_{N-1} \\ x_{N,d} \end{pmatrix}. \tag{5.103}$$

Now, taking the scalar product $E \cdot \dot{X}$ corresponds to inserting $\dot{X}$ into the RHS of (5.100) with $x_{N,d}$ substituted according to (5.95):

$$E \cdot \dot{X} = -k_0 \left[\left(\prod_{i=1}^{N-1} (D + k_i) \right) x_1 + \left(\prod_{i=0}^{N-1} (D + k_i) \right) (-x_0) \right] \tag{5.104}$$

Here, the first product takes care of the $x_1, ..., x_{N-1}$-terms, while the second product essentially provides $x_{N,d}$. Both products contribute one superfluous summand x_N each, which cancel each other out due to their opposite signs. Now, comparing (5.104) with the derivation of (5.99), and restricting X to any solution of (5.100), it follows that:

$$E \cdot \dot{X} = +k_0^2 \left(\prod_{i=1}^{N-1} (D + k_i) \right) x_0$$

$$= - \left(\prod_{i=0}^{N-1} k_i \right) (\pm \bar{x}_{1,d}) \tag{5.105}$$

Therefore, the system will evolve toward a smaller RHS of (5.100) from the boundary of positive traveling, and toward a greater RHS of (5.100) from the boundary of negative traveling. In other words, the position control phase (approach mode) can never be left toward traveling mode. As approach mode control is GAS according to proposition 4, solutions to (5.98) will be as well *after approach mode has been reached*.

Thus, the situation during traveling mode remains to be investigated. Again, $\dot{X}$ can be inserted into the RHS of (5.100), this time using (5.96,5.97):

$$E \cdot \dot{X} = -k_0 \left[\left(\prod_{i=1}^{N-1} (D + k_i) \right) x_1 + \left(\prod_{i=1}^{N-1} (D + k_i) \right) (\pm \bar{x}_{1,d} - x_1) \right]$$

$$= - \left(\prod_{i=0}^{N-1} k_i \right) (\pm \bar{x}_{1,d}) \tag{5.106}$$

Hence, the RHS of (5.100) will decrease at a constant rate in case of positive traveling, and increase at a constant rate in case of negative traveling. This implies that velocity control (traveling mode) will be replaced by position control (approach mode) after a finite time, and concludes the proof of proposition 9. □

Please note that when applied in a practical system, it is indeed possible that the control mode changes more than once in the vicinity of the boundary plane. This could very well result from noise and other disturbances in the real world. However, due to the smoothness of the controller output according to (5.98) and due to the consistent change rates in both (5.105) and (5.106), convergence to approach mode is still guaranteed, and the resulting control signal will not exhibit any undesirable artifacts.

Although not formalized here, mode-switching control according to (5.98) can be further generalized by delimiting the output of the traveling mode controller through an additional "acceleration" controller and a desired acceleration $\pm\bar{x}_{2,d}$. This becomes clear from the fact that proposition 9 can be recursively applied at the subsequent stage, using systems and controllers of orders $N-1$ and $N-2$, and so on if necessary. However, with more than two possible modes, it would be important to start the search for the "smallest" controller output at the Nth-order system and terminate the search as soon as one lower-order controller issues a "higher" output. Otherwise, undesirable jumps from a higher-order control mode back to a lower-order one could occur.

5.5   Example Controller: MARVIN

With the previous section, the set of formal tools for designing small UAV flight controllers has been completed. This section and the following two illustrate the application of these tools to the creation of actual controllers. The current section deals with the helicopter UAV MARVIN, developed in the author's group, the basic architecture of which has already been presented in section 2.3.3.

Section 5.5.1 below introduces the state description, section 5.5.2 exhibits the system dynamics underlying the controller design, and section 5.5.3 demonstrates how the model parameters can be identified. Then, section 5.5.4 presents the architecture of MARVIN flight control, instantiating figure 5.10. Section 5.5.5 discusses how the mode switching from section 5.4.5 almost automatically creates MARVIN's capability of flying multi-way-point trajectories. Finally, section 5.5.6 depicts the performance of the controller by means of data from real flight experiments.

5.5.1   State Description

This section introduces the state description used by the controller and points out the dynamical model equations with respect to the real and virtual control variables available.

The MARVIN flight controller is designed for roll and pitch angles not exceeding $\pm30°$. In this operating range, vertical and horizontal movements can be controlled independently with suitable decoupling transformations applied. Otherwise, if nearly-horizontal orientation of the rotor axis had been permitted, the vertical motion could *not* have been controlled in such flight situations, which in turns would have significantly complicated the controller and the generation of feasible trajectories.

For certain parts of the state description and model, special coordinate systems need to be introduced:

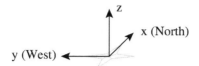

Figure 5.13: Base coordinate system (BCS).

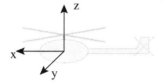

Figure 5.14: Vehicle coordinate system (VCS).

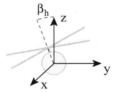

Figure 5.15: Force coordinate system (FCS).

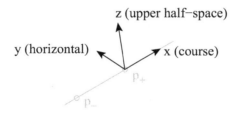

Figure 5.16: Segment coordinate system (SCS).

- The *base coordinate system* (BCS) is the global reference coordinate system, with x to geographic north and z upward. Its origin can be defined by the user, usually being some reference point within the current area of operation. See figure 5.13.

- The *vehicle coordinate system* (VCS) is the helicopter-fixed coordinate system, with x forward and z upward (parallel to the rotor axis). Its origin should reflect the helicopter's center of mass. See figure 5.14.

- The *force coordinate system* (FCS) is the VCS rotated about VCS-x by the angle $-\beta_h$ such that the resulting force F_R applied by the main and tail rotors of the helicopter is in the direction of FCS-z, or the unit vector in the direction of F_R is $\hat{F}_R = z_{FCS}$. β_h

can be observed as the helicopter's roll angle while hovering and amounts to about
4° with MARVIN. It is assumed that β_h is constant, which significantly simplifies the
dynamics model. Due to the above-mentioned restriction of the roll and pitch angles,
this assumption is very well justified, even more so when angular acceleration due to
tail rotor control is averaged out.

- The *segment coordinate system* (SCS) is defined according to the current desired course
  segment, which is always a straight line through two points, p_- and p_+. x is parallel to
  this course line, y is always horizontal (i.e. $y_{SCS} \cdot z_{BCS} = 0$), and z pointing toward the
  upper BCS half-space (i.e. $z_{SCS} \cdot z_{BCS} \geq 0$). Thus, if and only if the desired course line
  is vertical, SCS-z will also be horizontal. The origin of the SCS is p_+, expressing the
  current target point. See figure 5.16.

The SCS is an important design aspect of the controller that permits a convenient distinction between the traveling direction (SCS-x) and the lateral deviations from the course line
(SCS-y, SCS-z). This is particularly reasonable in combination with mode switching according to section 5.4.5, as will be explained in detail later. In the equations used in this and the
following subsections, above coordinate systems will mainly be used to represent orientation;
then, orthonormal matrices from $\mathbb{R}^3$ will be used to denote them. Vectors and matrices without
coordinate system specification will generally refer to the BCS.

The description of the system state, as used by MARVIN's flight controller, is parameterized
with the current SCS orientation and expressed as S_{SCS}:

$$S_{SCS} = \begin{pmatrix} {}^{SCS}p \\ {}^{SCS}v \\ z_{FCS}^x \\ z_{FCS}^y \\ \psi \\ \omega_R \end{pmatrix} \in \mathbb{R}^{10} \tag{5.107}$$

Here, $SCS \in \mathbb{R}^{3\times3}$ is the orthonormal matrix denoting the orientation of the SCS relative to the
BCS. This orientation is assumed to be *constant*. That is, switching to a new course segment
is regarded as initiating a new convergence process of the closed-loop system.

The components of S_{SCS} denote:

- ${}^{SCS}p = {}^{SCS}BCS \cdot (p - p_+) \in \mathbb{R}^3$ is the helicopter's BCS position p, but expressed in the
  SCS and relative to the current target way-point p_+.

- ${}^{SCS}v = {}^{SCS}\dot{p} \in \mathbb{R}^3$ is the helicopter's velocity in SCS coordinates.

- $z_{FCS}^x, z_{FCS}^y \in \mathbb{R}$ are the BCS-x,y coordinates of the unit vector $\hat{F}_R$ in the direction of the
  current resulting rotor force.

- $\psi \in [0; 2\pi)$ is the "standard" yaw Euler angle of the helicopter, given that the VCS
  orientation can be expressed through Euler angles (ψ, θ, ϕ) for yaw, pitch, and roll,
  respectively, as:

$$^{BCS}VCS = \mathrm{Rot}_z(\psi) \cdot \mathrm{Rot}_y(\theta) \cdot \mathrm{Rot}_x(\phi) \tag{5.108}$$

Hence, ψ is the angle between BCS-x ("north") and the BCS-horizontal projection of VCS-x ("forward"), which equals FCS-x. (5.108) can be unambiguously inverted for any pitch angle $\|\theta\| < \pi/2$ yielding:

$$
\begin{pmatrix} \psi \\ \theta \\ \phi \end{pmatrix} = \begin{pmatrix} \mathrm{atan2}\left(x^y_{VCS}, x^x_{VCS}\right) \\ \mathrm{atan2}\left(-x^z_{VCS}, \sqrt{(x^x_{VCS})^2 + (x^y_{VCS})^2}\right) \\ \mathrm{atan2}\left(y^z_{VCS}, z^z_{VCS}\right) \end{pmatrix} \tag{5.109}
$$

- $\omega_R \in \mathbb{R}$ is the current rotation rate of the main rotor, relative to the VCS (i.e. as read by an RPM sensor mounted on board).

This state description is actually minimal with respect to the number of scalar components, albeit not intuitive in the orientation coordinates chosen. However, this choice already reflects the intended controller design and will prove to ideally fit the required system model structure according to figure 5.10 and proposition 4.

5.5.2 System Dynamics

For the dynamics model, the first six system equations are:

$$
\begin{align}
{}^{SCS}\dot{p}^x &= {}^{SCS}v^x \tag{5.110} \\
{}^{SCS}\dot{p}^y &= {}^{SCS}v^y \tag{5.111} \\
{}^{SCS}\dot{p}^z &= {}^{SCS}v^z \tag{5.112} \\
{}^{SCS}\dot{v}^x &= (1/m_H) \cdot {}^{SCS}F^x \tag{5.113} \\
{}^{SCS}\dot{v}^y &= (1/m_H) \cdot {}^{SCS}F^y \tag{5.114} \\
{}^{SCS}\dot{v}^z &= (1/m_H) \cdot {}^{SCS}F^z \tag{5.115}
\end{align}
$$

These are trivial laws of motion, with (5.113–5.115) stating Newton's law with some cumulative force vector ${}^{SCS}F$ (which includes gravity) and m_H denoting the mass of the helicopter. These equations certainly hold with zero error. ${}^{SCS}F$ is a virtual control in the respect of being actually exerted through the remaining stages of the system. It is related to the force F_R exerted by the two rotors via

$$
\begin{align}
{}^{SCS}F &= {}^{BCS}SCS^T \cdot \left(F_R + (0,0,-m_H \cdot g)^T\right) \tag{5.116} \\
F_R &= {}^{BCS}SCS \cdot {}^{SCS}F + (0,0,m_H \cdot g)^T \tag{5.117}
\end{align}
$$

with g expressing the magnitude of gravity. The last two equations constitute a transform function and its inverse, respectively, according to figure 5.6.

Using the components z^x_{FCS}, z^y_{FCS} from S_{SCS}, the model equation for F_R is

$$
F_R = \|F_R\| \cdot \left(z^x_{FCS}, z^y_{FCS}, \sqrt{1 - \left(z^x_{FCS}\right)^2 - \left(z^y_{FCS}\right)^2}\right)^T \tag{5.118}
$$

with $\|F_R\|$ being provided through collective pitch control input, see below. This transform stage is continuously invertible as well.

Now, the change rates of z^x_{FCS}, z^y_{FCS}, ψ need to be written down as a transform stage model equation. It is

$$
\begin{aligned}
\dot{z}_{FCS} &= \omega \times z_{FCS} \\
&= \left(^{BCS}FCS \cdot {}^{FCS}\omega\right) \times z_{FCS} \\
&= \left(^{FCS}\omega^x \cdot x_{FCS} + {}^{FCS}\omega^y \cdot y_{FCS} + {}^{FCS}\omega^z \cdot z_{FCS}\right) \times z_{FCS} \\
&= -{}^{FCS}\omega^x \cdot y_{FCS} + {}^{FCS}\omega^y \cdot x_{FCS}
\end{aligned}
\tag{5.119}
$$

or as system of equations for each component:

$$
\dot{z}^x_{FCS} = {}^{FCS}\omega^y \cdot x^x_{FCS} - {}^{FCS}\omega^x \cdot y^x_{FCS} \tag{5.120}
$$
$$
\dot{z}^y_{FCS} = {}^{FCS}\omega^y \cdot x^y_{FCS} - {}^{FCS}\omega^x \cdot y^y_{FCS} \tag{5.121}
$$

The change rate ψ of the yaw angle, in turn, can be calculated by

$$
\psi = \frac{\dot{x}_{FCS} \cdot \hat{\psi}_n}{r} \tag{5.122}
$$

with $\hat{\psi}_n$ denoting the unit normal vector of the current "yaw plane", i.e. the vertical BCS plane containing x_{FCS}, and r denoting the radius of the endpoint of x_{FCS} while rotating about BCS-z. Hence,

$$
r = \sqrt{\left(x^x_{FCS}\right)^2 + \left(x^y_{FCS}\right)^2} \tag{5.123}
$$
$$
\hat{\psi}_n = \begin{pmatrix} -x^y_{FCS} \\ +x^x_{FCS} \\ 0 \end{pmatrix} \cdot \frac{1}{r}. \tag{5.124}
$$

$\dot{x}_{FCS}$ is most easily obtained through its representation in the FCS and subsequent transformation into the BCS:

$$
\begin{aligned}
\dot{x}_{FCS} &= {}^{BCS}FCS \cdot \begin{pmatrix} 0 \\ +{}^{FCS}\omega^z \\ -{}^{FCS}\omega^y \end{pmatrix} \\
&= {}^{FCS}\omega^z \cdot y_{FCS} - {}^{FCS}\omega^y \cdot z_{FCS}
\end{aligned}
\tag{5.125}
$$

Now, the last three findings can be inserted into (5.122), yielding:

$$
\psi = \frac{{}^{FCS}\omega^z \cdot \left(y^y_{FCS}x^x_{FCS} - y^x_{FCS}x^y_{FCS}\right) + {}^{FCS}\omega^y \cdot \left(z^x_{FCS}x^y_{FCS} - z^y_{FCS}x^x_{FCS}\right)}{\left(x^x_{FCS}\right)^2 + \left(x^y_{FCS}\right)^2} \tag{5.126}
$$

The system of equations (5.120,5.121,5.126) is continuously invertible for $^{FCS}\omega$ provided that the horizontal projections of x_{FCS} and y_{FCS} are linearly independent. This is the case for FCS roll and pitch angles below $\pm 90°$, certainly so for helicopter roll and pitch angles within $\pm 30°$

as specified for the MARVIN controller. Using FCS-rotations in this part of the model is only one of many choices formally, but the fact that $\dot{z}_{FCS}$ does not depend on ω_z and ψ does not depend on ω_x in the FCS yields a system of equations which is particularly simple to invert.

As the cyclic and tail rotor pitch control inputs act relative to the VCS instead of the FCS, another simple coordinate transform needs to be applied beforehand to have (5.120,5.121,5.126) depend on $^{VCS}\omega$. With

$$^{FCS}VCS = \text{Rot}_x(\beta_h) = \begin{pmatrix} 1 & 0 & 0 \\ 0 & \cos\beta_h & -\sin\beta_h \\ 0 & \sin\beta_h & \cos\beta_h \end{pmatrix}, \tag{5.127}$$

it is:

$$^{FCS}\omega = {}^{FCS}VCS \cdot {}^{VCS}\omega \tag{5.128}$$

$$^{VCS}\omega = {}^{FCS}VCS^{\text{T}} \cdot {}^{FCS}\omega \tag{5.129}$$

One additional remark refers to the fact that ^{BCS}FCS, which is used in the rotation-dependent dynamics, does not actually belong to the state variables of the system. Therefore, one should be aware that – and how – it can be deduced from available "fed-back" state variables at any time, consistent with (5.7):

$$z_{FCS} = \left(z_{FCS}^x, z_{FCS}^y, \sqrt{1 - \left(z_{FCS}^x \right)^2 - \left(z_{FCS}^y \right)^2} \right)^{\text{T}} \tag{5.130}$$

$$x_{FCS} = (-\sin\psi, \cos\psi, 0)^{\text{T}} \times z_{FCS} \tag{5.131}$$

$$y_{FCS} = z_{FCS} \times x_{FCS} \tag{5.132}$$

Up to this point, only trivial motion dynamics and coordinate transforms, without any relation to helicopter physics, have been used in the model equations. The remaining model equations, though, will refer to the control inputs and their effects as known from helicopter theory. The latter has been extensively addressed in the literature, for example in [36], as well as in a recent PhD thesis [38] created in the author's group in the context of the MARVIN projects.

The force F_R applied by the rotors of the helicopter depends linearly on the main rotor collective pitch control input P_c. Following the FCS convention as introduced in section 5.5.1, this means:

$$^{FCS}F_R = (0, 0, P_c/f_{Pc})^{\text{T}} \tag{5.133}$$

The proportionality between collective pitch and rotor force within the desired operating range is one key property of helicopter dynamics. The factor of proportionality f_{Pc} needs to be determined (see 5.5.3 how).

For the cyclic pitch control inputs P_x and P_y, due to gyroscopic effects of the main rotor, the following law is a very good approximation of the actual dynamics [71]:

$$^{VCS}\omega^x = (1/f_{Pxy}) \cdot P_x \tag{5.134}$$

$$^{VCS}\omega^y = (1/f_{Pxy}) \cdot P_y \tag{5.135}$$

This direct dependence on the control inputs clarifies why the rotation rate vector $^{VCS}\omega$ does not belong to the state variables in S_{SCS}. Of course, the omission of the laws of accelerated rotation constitutes a certain simplification. But as this simplification is quantitatively in the order of magnitude of 1 % [38][3], it will be clearly dominated by several kinds of unavoidable disturbances and uncertainties, anyway.

As the tail rotor control input P_z is fed through a commercial gyroscope module, the respective dynamics, too, observe a very good proportionality between P_z and the resulting VCS-z rotation rate [84]:

$$^{VCS}\omega^z = (1/f_{Pz}) \cdot P_z \qquad (5.136)$$

Finally, the change rate of the main rotor RPM ω_R can reasonably be seen as proportional to the throttle control input th, but it also depends on changes in the main rotor torque due to the current collective pitch setting P_c [38]:

$$\dot{\omega}_R = (1/f_{th}) \cdot th - M_{th} \cdot P_c^2 \qquad (5.137)$$

Here, M_{th} is a scaling factor depending on the rotor geometry. While this proportionality between $\dot{\omega}_R$ and th is not as good over the *possible* range of ω_R as the previous dynamics equations, it is fully sufficient especially since the desired RPM is always constant in flight. Thus, the resulting error is very small although (5.137) may be considered an operating-point linearization.

This already concludes the examination of the open-loop system model. Please note that all laws describing the direct effects of control inputs are simple linear relations, while all nonlinearities are just coordinate transforms and obey the requirement of continuous invertibility according to (5.9).

5.5.3   Parameter Identification

It remains to be seen how the parameters in the linear dynamics equations, i.e. f_{Pc}, f_{Pxy}, f_{Pz}, f_{th}, can be identified. This is done through real-flight experiments under human remote control with data-logging. Figures 5.17–5.19 show real-flight plots of some control inputs together with sensor readings that linearly depend on them. Whenever the controller is set up for a new helicopter, an "identification flight" is performed to determine the f-factors mentioned above. For this, a comparison of the two amplitudes in each of the plots suffices – the bias will later be set automatically by a bias compensation stage according to figure 5.8. The plots also confirm the good validity of the postulated linearities. While cyclic pitch (figure 5.17) and tail rotor output (figure 5.18) are proportional to one of the measured VCS rotation rates, collective pitch (figure 5.19) is plotted against the measured VCS-z acceleration. The latter measurement is a little noisier than the former ones, which explains the slightly lower quality of the comparison. Please note the extreme similarity between the two graphs in figure 5.18, given that the commercial gyroscope module is also involved there!

---

[3]Physically, the ratio of the angular momentum of the main rotor's rotation on the one hand and the momentum of the helicopter's roll and pitch rotation on the other is the key to this finding. The momentum of the roll and pitch rotation can effectively be neglected.

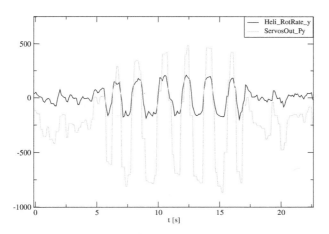

Figure 5.17: Proportionality between pitching rotation rate ($1024 \hat{=} 90°/s$) and corresponding cyclic pitch of the main rotor (in servo ticks).

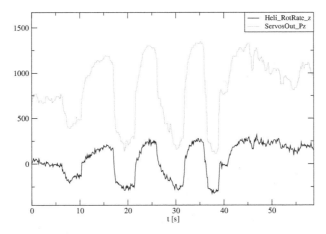

Figure 5.18: Proportionality between vertical rotation rate ($1024 \hat{=} 90°/s$) and tail rotor gyroscope input (in servo ticks).

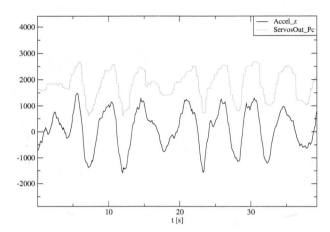

Figure 5.19: Proportionality between vertical acceleration (in mm/s^2) and collective pitch of the main rotor (in servo ticks).

5.5.4   Control Architecture

Figure 5.20 presents the functional architecture of the MARVIN flight controller. This section explains the architecture and connects it with the dynamics model of section 5.5.2 and the generic controller construction process according to section 5.3.

In the diagram, the boxes labeled *C0/1*, *C1/2*, or *C1* refer to (cascades of) integrator stage controllers with the specified order N according to figure 5.5 and (5.13). Particularly, *C1/2*, for example, denotes a mode-switching controller of variable order $N = 1..2$ according to (5.98). *BC* marks bias compensation stages according to figure 5.8, and *BC** refers to "smart" bias compensation according to figure 5.9, with the required corresponding downstream projections (plant-wise) indicated through dashed arrows. All remaining blocks, painted in grey, denote calculations that effectively constitute inverted transform functions f_j^{-1} (e.g. coordinate transforms) in the sense of figure 5.6.

In the top left region, SCS-position and -velocity are fed through three mode-switching controllers *C1/2* controlling the integrators according to (5.110–5.115). The controllers output the desired SCS acceleration vector $^{SCS}a_d$. The mode-switching feature of these controllers is the principal reason for the use of the SCS in general: Usually, only the SCS-x motion controller should apply velocity control at greater distances from the target, while the y and z controllers just keep the helicopter on the course segment line through position control. The definition of the SCS helps to keep this control regime independent of the actual traveling direction.

In the next step, the desired acceleration is transformed into BCS representation, yielding a_d. This transform corresponds to that in (5.117) in the model. In order to obtain $F_{R,d}$, the desired rotor force in the BCS, gravity needs to be added in positive z direction and the acceleration signal scaled with the mass m_H of the helicopter. However, this step is augmented with a bias compensation stage in x and y directions. The main reason for a bias in the model at this stage

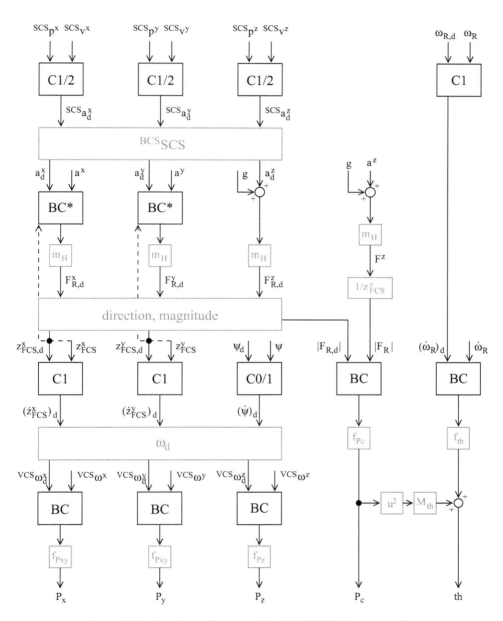

Figure 5.20: Functional diagram of the MARVIN flight controller.

is atmospheric wind, the effect of which is most reasonably compensated by BCS-anchored integrators – for the wind would not, for example, turn together with the helicopter[4]. This bias compensation stage is executed as "smart" bias compensation according to figure 5.9, since it is separated from the plant's control inputs through another level of controllers.

Now, $F_{R,d}$ is functionally transformed into z^x_{FCS}, z^y_{FCS}, and $\|F_{R,d}\|$, inverting (5.118). The desired change rates $(\dot{z}^x_{FCS})_d$ and $(\dot{z}^y_{FCS})_d$ are computed by two standard integrator-stage controllers $(C1)$ according to (5.13). Furthermore, the yaw angle controller computes the desired change rate of the yaw angle $(\dot{\psi})_d$ from the current error $\psi_d - \psi$. This controller, labeled $C0/1$, implements a mode-switching controller according to (5.98) that just outputs a constant desired change rate $\pm\bar{\omega}$ in the $N = 0$ case of "velocity control". The desired value ψ_d is constant for every course line segment and corresponds to the projection of x_{SCS} into the horizontal plane of the BCS[5]:

$$\psi_d \;=\; \mathrm{atan2}\left(x^y_{SCS}, x^x_{SCS}\right) \tag{5.138}$$

In section 5.5.5 about multi-point trajectories, a different case for the determination of the desired heading will be dealt with, which is curve flying.

With $(\dot{z}^x_{FCS})_d$, $(\dot{z}^y_{FCS})_d$, and $(\dot{\psi})_d$ calculated, the transform function block labeled ω_d computes the desired rotation rates in the VCS. Effectively, this means first inverting (5.120,5.121) to yield

$$^{FCS}\omega^x \;=\; \frac{\dot{z}^x_{FCS}x^y_{FCS} - \dot{z}^y_{FCS}x^x_{FCS}}{y^y_{FCS}x^x_{FCS} - y^x_{FCS}x^y_{FCS}} \tag{5.139}$$

$$^{FCS}\omega^y \;=\; \frac{\dot{z}^x_{FCS}y^y_{FCS} - \dot{z}^y_{FCS}y^x_{FCS}}{y^y_{FCS}x^x_{FCS} - y^x_{FCS}x^y_{FCS}}, \tag{5.140}$$

which is always solvable due to the limitation of the pitch and roll angles. Then, $^{FCS}\omega_y$ from (5.140) is inserted into (5.126) yielding:

$$^{FCS}\omega_z \;=\; \frac{\dot{\psi}\left(\left(x^x_{FCS}\right)^2 + \left(x^y_{FCS}\right)^2\right) - {}^{FCS}\omega^y\cdot\left(z^x_{FCS}x^y_{FCS} - z^y_{FCS}x^x_{FCS}\right)}{y^y_{FCS}x^x_{FCS} - y^x_{FCS}x^y_{FCS}} \tag{5.141}$$

Using $(\dot{z}^x_{FCS})_d$, $(\dot{z}^y_{FCS})_d$, and $(\dot{\psi})_d$ for the respective variables in (5.139–5.141) gives $^{FCS}\omega_d$, and finally applying (5.129) results in

$$^{VCS}\omega_d \;=\; {}^{FCS}VCS^T \cdot {}^{FCS}\omega_d, \tag{5.142}$$

which is the result of this transformation block in the controller.

At last, one bias compensation stage is applied to each of the desired VCS rotation rates, and scaling with the factors of proportionality according to (5.134–5.136) yields the control inputs P_x, P_y, and P_z fed to the cyclic pitch servos and the tail rotor gyroscope module.

---

[4]The precise way of introducing bias compensation stages into a controller is always somewhat arbitrary, from a formal point of view. Yet, considering the nature of the dominating disturbances as in this example will help in optimizing the quantitative performance of the controller.

[5]Whenever x_{SCS} is vertical, ψ_d is kept from the last course segment for which it was not.

The desired magnitude $\|F_{R,d}\|$ of the rotor force can directly be set via the collective pitch control input P_c as stated in (5.133). But it is required to provide a bias compensation stage first, because disturbances like atmospheric wind and the ground effect significantly change the induced velocity of the airflow through the rotor disc and, as a result, $\|F_R\|$ [38, 90, 80]. The bias compensation stage to be applied here needs the actual $\|F_R\|$ value as its second input, which is estimated from the vertical BCS acceleration a^z and the responsible force F^z through:

$$F^z = m_H \cdot (a^z + g) = \|F_R\| \cdot z^z_{FCS} \tag{5.143}$$

$$\|F_R\| = \frac{m_H \cdot (a^z + g)}{z^z_{FCS}} \tag{5.144}$$

Due to the postulated limitation of the roll and pitch angles, z_{FCS} is always large enough for this to work properly. After the bias compensator, scaling with f_{Pc} provides the control input P_c.

RPM (throttle) control is a straightforward single-integrator and bias compensation type control sequence. The desired RPM $\omega_{R,d}$ is always constant in flight, around 1300min^{-1} for MARVIN. Inverting (5.137), it is

$$th = f_{th} \cdot \left((\dot{\omega}_R)_d + M_{th} \cdot P_c^2 \right), \tag{5.145}$$

which completes the computation of control inputs.

5.5.5 Multi-Point Trajectories

In the previous sections, only single target points have been addressed. This section explains how the MARVIN flight controller has been added the capability of handling way-point lists defining a series of linear course segments that result in multiple-way-point trajectories without stopping at any intermediate way-point. Subsequent linear course segments in such a list are connecting by flying a *curve*.

Curve-Flight at Segment Transition

Let $\langle p_i \rangle$ be the list of way-points. When the nth course segment is the currently active one, it is

$$p_-(t) = p_{n-1} \tag{5.146}$$

$$p_+(t) = p_n \tag{5.147}$$

as described in section 5.5.1 for defining the segment coordinate system SCS. This segment is used until the SCS motion controllers (yielding a_d) have mode-switched to position control according to (5.98). At that time, flight control automatically advances to the subsequent segment in the list, setting

$$p_-(t) = p_n \tag{5.148}$$

$$p_+(t) = p_{n+1} \tag{5.149}$$

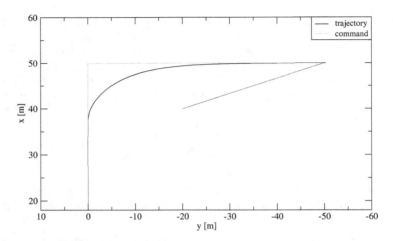

Figure 5.21: Simulated trajectory for a multi-way-point flight path ($k_1 = k_2 = 0.3\,\mathrm{s}^{-1}$, $r_v = 2\,\mathrm{m/s}$)

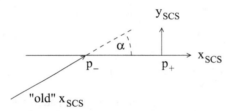

Figure 5.22: Transition to new SCS and curve flying.

provided the end of the list is not yet reached. Then, the "new" SCS-y and -z controls guide the helicopter onto the new course segment laterally while the new SCS-x control accelerates to reach the nominal traveling speed $\bar{v}_d$ according to (5.98). The result is a *curve*.

Figure 5.21 shows an example of a resulting flight trajectory. In this example, the helicopter starts from the bottom left corner flying forward along the x-axis and then starts a curve turning right by 90°. One can see that the straight line along the x-axis is left at $x \approx 37.7$ m. From now on the next point in the list is used as p_+, making $p_+^x = 50$ m, $p_+^y = -50$ m. The exponential convergence to $x = 50$ m is clearly visible in the figure.

To analyze the control behavior when advancing to a new course segment more closely, one may solve (5.95–5.98) for the time of switching, i.e. postulate $a_d^+(t) = a_d(t)$ using the naming conventions from the controller diagram:

$$k_1(\bar{v}_d - v) = k_0 k_1 (p_d - p) - (k_0 + k_1)v \tag{5.150}$$

$$p_d - p = \frac{1}{k_0}\bar{v}_d + \frac{1}{k_1}v \tag{5.151}$$

Assume that the flight trajectory had already converged to the course line, i.e. zero error in the old SCS-y and -z directions, so that all non-zero control according to (5.150) is related to the old SCS-x direction. Figure 5.22 depicts the new SCS in its xy-plane and the projection of the old SCS-x traveling direction into this plane. Then, the two state variables $(p_d - p)$ and v scale with $\sin\alpha$ to yield the new SCS-y control (examining the new SCS-z would be fully analog). First, it is clear that the new SCS-y controller applies position control from the start, because $\bar{v}_d$ is *not* scaled with $\sin\alpha$ but remains unchanged. So the RHS of (5.150) scaled with $\sin\alpha$ constitutes the new desired SCS-y acceleration, which is therefore equal to the former desired SCS-x acceleration projected onto the new SCS-y. Summarizing, the only change (i.e. non-continuity) of the desired acceleration vector a_d occurs in the new SCS-x direction, i.e. the new direction of traveling. This is actually the most desirable behavior.

In the special case of an approach with the desired velocity $v = \bar{v}_d$, (5.150) simplifies to

$$\frac{p_d - p}{\bar{v}_d} = \frac{1}{k_0} + \frac{1}{k_1} = T \tag{5.152}$$
$$p_d - p = T \cdot \bar{v}_d \tag{5.153}$$

with T denoting the "time to impact" at the target location assuming constant velocity.

While a "curve" is flown by the helicopter after advancing to a new course segment, the current horizontal projection of the *actual* velocity vector is utilized as the desired yaw angle, in order to have the helicopter heading "forward" (this is also beneficial for aerodynamical reasons at higher traveling velocities). Thus, (5.138) is replaced during curve flying by:

$$\psi_d = \operatorname{atan2}(v^y, v^x) \tag{5.154}$$

In order to limit the change rate of ψ_d resulting from (5.154), curve flying is subject to two additional conditions. All in all, the transition to the next course segment in "curve mode" happens if and only if:

1. all three *C1/2* SCS position controllers (a_d^x, a_d^y, a_d^z) are in position-control mode, and

2. the angle between the horizontal projections of the current velocity v and the new x_{SCS} is below α_0, and

3. the magnitude of the current actual velocity is $\|v\| > v_0$,

with α_0, v_0 denoting two parameters, which are $\alpha_0 = 120°$ and $v_0 = 1$ m/s in the case of MARVIN. Here, conditions 2 and 3 assure that ψ_d cannot change unboundedly fast during the curve.

The effect of this rule is also depicted in figure 5.21: The curve to the right is entered according to above conditions. But there is another course segment from $p_-^x = 50$ m, $p_-^y = -50$ m to $p_+^x = 40$ m, $p_+^y = -20$ m, which cannot be entered via a curve due to condition 3. Instead, the helicopter approaches the old p_+ (i.e. new p_-) until its velocity is nearly zero, turns into the next segment's direction while hovering, and follows this next segment then. Therefore, the actually flown trajectory is exactly congruent with the final desired course segment in the figure.

The formal precondition for the transition to the next segment *without* flying a curve is the conjunction of above condition 1 and the negation of condition 3. This latter part expresses that the helicopter has "nearly" come to hovering at the former target location.

Operation Space Estimation

Follow-up sequences of course segments will never be followed exactly, of course. This section presents a way to estimate a subspace that will contain the resulting trajectory except for very small errors. This space is called *operation space*.

Assume that the helicopter is initially hovering at the first course segment's starting point p_0, $\bar{v}_d$ is the desired traveling speed, and the way-point-list consists of $N+1$ points $p_0...p_N$ (i.e. N segments). Then the following algorithm calculates the operation space $O \subset \mathbb{R}^3$ for following this list of way-points as described in the previous subsection as the sum of N segment-wise operation spaces $o_j \subset \mathbb{R}^3$:

1. The first segment is equal to its operation space, i.e. $o_1 = \overline{p_0 p_1}$.

2. For each corner point p_j between two adjacent segments $(0 < j < N)$, calculate the intersection of the previous segment's operation space with the sphere $\|r - p_j\| \leq T \cdot \bar{v}_d$ around p_j with T from (5.152). This intersection is called the subsequent segment's *entrance space* e_{j+1}.

3. The subsequent segment's operation space o_{j+1} is the convex hull of its entrance space e_{j+1} and its target point p_{j+1}.

4. The resulting operation space is the sum of all segments' operation spaces:

$$ O = \bigcup_{i=1}^{N} o_i \tag{5.155} $$

This construction algorithm is slightly idealized and will not be examined formally here. However, it is based on the following formal aspects:

- (5.151) basically describes a fixed distance $\bar{v}_d/k_0$ around any target point p_j which is expanded or reduced according to the current v/k_1 in order to decide the space subject to position control. It follows from above condition 1 and proposition 9 that the distance $\bar{v}_d/k_0$ cannot be left by the helicopter after p_j has become the starting point of the current course segment.

- The time of switching assures that no target overshooting will occur in any one of the three new SCS directions – this follows from figure 5.22 and proposition 8.

- The surfaces of the convex hulls used in building the operation spaces o_j consist of trajectories resulting from $v = 0$ initially and position control in all three SCS coordinates. Using velocity control for SCS-x will lead to relatively *faster* lateral convergence.

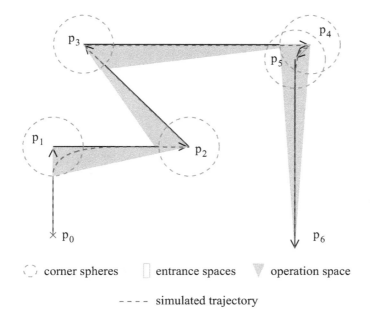

- ◌ corner spheres   ▯ entrance spaces   ▽ operation space

- - - - simulated trajectory

Figure 5.23: Sample calculation of the operation area and simulated flight trajectory.

Figure 5.23 depicts this algorithm for a sample sequence that is two-dimensional and shows the resulting actual trajectory from a simulated flight. This trajectory was generated using $\bar{v}_d = 3$ m/s, p_4 is at position $x = 170$ m, $y = 160$ m relative to the starting point p_0. The spheres have a radius r of

$$r = T \cdot \bar{v}_d = 3\,\text{m/s} \cdot 2 \cdot \frac{1}{0.3\,\text{s}^{-1}} = 20\,\text{m}. \qquad (5.156)$$

5.5.6 Experimental Results

The controller architecture described in 5.5.4 performs well in stabilizing the MARVIN helicopter at a single point as well as in flying complete missions. This section documents the performance of the controller by means of real outdoor experiments with MARVIN.

Figure 5.24 depicts a trajectory of a flight experiment. Starting from a position at $x = 45$ m, the helicopter went to the destination position at $x = 90$ m. This is a flight along the x-axis of the BCS. The corresponding velocity is scaled by a factor of 10 for better legibility.

One can see the exponential convergence of the position to the desired value. The velocity curve shows the acceleration phase with an exponential convergence of the velocity to the given maximum speed. During this phase until $t \approx 25$ s, velocity control is active. Then the controller switches to position control mode so that the remaining position error as well as the velocity converge to zero.

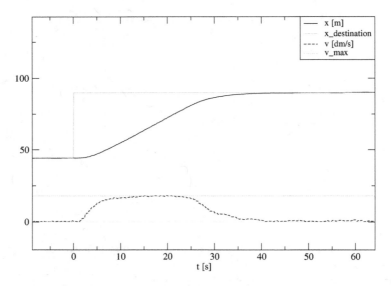

Figure 5.24: Trajectory flown in experiment with MARVIN.

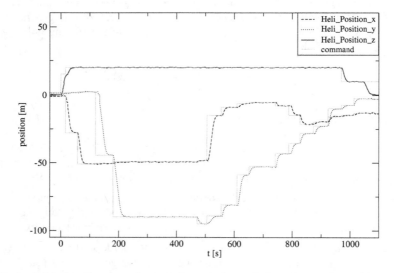

Figure 5.25: Separate coordinates of a complete mission, flown with MARVIN.

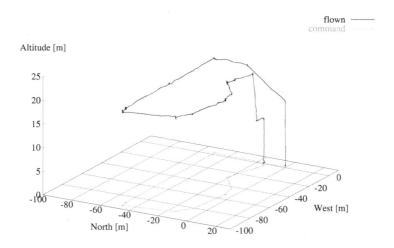

Figure 5.26: 3D representation of a complete mission, flown with MARVIN.

In complete missions, such approach phases to new way-points are performed multiple times. Figure 5.25 shows a mission of almost 20 minutes including an autonomous take-off and landing. One can see many convergence trajectories in all three axes. While the altitude is constant at 20 m for nearly the complete flight, the horizontal position is subject to many changes.

Between $t = 200$ s and $t = 470$ s MARVIN hovered at one point for monitoring a point of interest with its on-board camera. This nicely shows the accuracy of position control.

In figure 5.26 the same flight is depicted in 3D representation, together with its projection onto the ground. Take-off took place at the rightmost point at $(x,y)^T = (0\,\text{m},0\,\text{m})^T$. The long hovering took place at the leftmost corner.

Such a mission is managed by a *mission control* entity, which provides the flight controller with the coordinates of way-points. Right above take-off one can see a significant deviation between the command and the trajectory actually flown. The initial destination point for take-off is given exactly above the helicopter. In this flight the subsequent point was sent to the controller before the helicopter had actually reached the initial destination altitude. Thus the controller immediately used the new course segment.

For the demonstration of the performance of the controller regarding multi-point trajectories according to 5.5.5, several experiments flying the "figure eight" have been performed with MARVIN. The result of such an experiment is shown in figure 5.27. The given way-points are shown together with the flown trajectory.

One can see the take-off and landing positions at the top right. The flight began in the center of the "eight" and with the lower circle. Since this circle was flown twice, the good repeatability of the trajectory is expressed. Although curve-flying according to section 5.5.5 does not intend to fly properly circular arcs in any way, the octagons of course segments used here lead to a

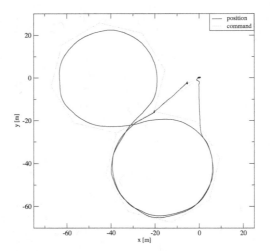

Figure 5.27: Figure-eight-shaped trajectory, flown in experiment with MARVIN.

nearly-circular trajectory.

5.6 Example Controller: Airship UAV

In this section, a controller design according to 5.3 for flight control of an airship UAV is presented. This is mainly indented as a practical demonstration of the generality of the design approach depicted in this chapter, rather than as the documentation of a full controller development process. Therefore, no experiments on the quantitative performance of this controller, neither simulated nor practical, will be presented.

Following roughly the same scheme as used in section 5.5, subsection 5.6.1 presents the assumed plant model, while subsection 5.6.2 explains the control architecture.

5.6.1 System Model

The airship assumed here exhibits only the following three control inputs (see section 1.3.1):

- th, the engine control signal ("throttle"),

- P_z, the pitch setting of the yaw rudder, and

- P_y, the pitch setting of the elevator.

The state description is restricted to the following components and, analog to the MARVIN model, parameterized by the current course segment coordinate system SCS:

$$S_{SCS} = \begin{pmatrix} {}^{SCS}p \\ \psi \\ \theta \end{pmatrix} \in \mathbb{R}^5 \tag{5.157}$$

The components of S_{SCS} denote:

- ${}^{SCS}p = {}^{SCS}BCS \cdot (p - p_+) \in \mathbb{R}^3$ is the airship's SCS position (i.e. position error), just as defined in (5.107) for MARVIN.

- $\psi, \theta \in (-\pi; \pi]$ are the "standard" yaw and pitch Euler angles of the airship, corresponding to (5.108). Hence, ψ is the angle between BCS-x ("north") and the BCS-horizontal projection of the airship's heading, while θ is the angle between the airship's heading and the BCS-horizontal plane.

The reader will most probably miss state variables expressing the current velocity. However, the airship dynamics model used here addresses an "equilibrium" subclass of states only, in the sense that the velocity vector is *already* parallel to the heading vector (x_{VCS} in the MARVIN convention). Of course, this is only reasonable in a system with zero atmospheric wind. This leads to the introduction of a new coordinate system, the *WCS*, which is oriented like the BCS but always moving at the current atmospheric wind-speed. Hence, the airship model assumes

$$^{WCS}v \parallel x_{VCS} = \text{Rot}_z(\psi) \cdot \text{Rot}_y(\theta) \cdot (1,0,0)^{\text{T}}. \tag{5.158}$$

Of course, this view of dynamics is simplified. On the other hand, due to the size-thrust-ratio typical for airships, the aerodynamic effect of the large tail fins usually present, and the typical slowness of maneuvering, the error resulting from this simplification should be sufficiently small, as long as the magnitude $\|^{WCS}v\|$ of the airspeed exceeds a certain threshold. This requirement will be explicitly addressed by the controller (see 5.6.2 below). Furthermore, all controller gains need to be set low enough to keep the quantitative error of the "equilibrium" assumption acceptably bounded.

All in all, the model presumes something like "roller-coaster kinematics", with the air working like rails. The resulting controller might not reach the maneuvering accuracy observed in MARVIN's flight control, but this is neither required for – nor can be expected from – the operation of an airship, anyway.

Now, the model equations will be presented, starting at the control inputs. For the throttle setting th, assuming that it is proportional to the voltage V applied to one or more electric motors, the basic dependencies observe

$$U \sim \omega \sim \sqrt{F} \sim v^* \tag{5.159}$$

with ω denoting propeller RPM, F denoting propulsion force, and v^* expressing the magnitude of the "equilibrium" speed, i.e. $v^* = \|^{WCS}v\|$. Therefore, the dynamics law for throttle is

$$v^* = (1/f_{th}) \cdot th \tag{5.160}$$

with the parameter f_{th} representing the factor of proportionality.

For the yaw and pitch rudders, a good approximation of the dynamics is

$$\dot{\psi} = (v^*/f_r) \cdot P_z \qquad (5.161)$$
$$\dot{\theta} = (v^*/f_r) \cdot P_y \qquad (5.162)$$

with the parameter f_r representing a factor of proportionality. These two equations result from the following consideration:

- In the equilibrium case, a constant P_z leads to a (slightly) changed lateral angle of attack α of the air to the body of the airship[6]. It is assumed that $\alpha \sim P_z$.

- For the centripetal force F_c resulting from α holds $F_c \sim (v^*)^2 \cdot \alpha$.

- F_c enforces a circular motion that rotates the velocity the airship's body – together with its velocity vector – at an angular rate of $\dot{\psi}$ such that $\dot{\psi} \cdot v^* \sim F_c$.

All in all, above relations combine into

$$P_z \;\sim\; \alpha \;\sim\; F_c/(v^*)^2 \;\sim\; \dot{\psi} \cdot v^*/(v^*)^2 \;\sim\; \dot{\psi}/v^* \qquad (5.163)$$

which corresponds to (5.161,5.162). Please note that this derivation assumes relatively low pitch angles, say below $\pm 30°$, so that (5.161) can reasonably be seen as independent of θ. Furthermore, the model assumes that the airship's roll angle is passively stabilized at $\phi = 0$ through the mass distribution within the vehicle, requiring no intervention at all by the controller. On the other hand, this mass distribution will actually interfere with (5.162), which is also neglected here considering "small" pitch angles.

Now, from ϕ, θ, v^* the derivation of ^{WCS}v is trivial:

$$^{WCS}v \;=\; \text{Rot}_z(\psi) \cdot \text{Rot}_y(\theta) \cdot (v^*, 0, 0)^\mathsf{T} \qquad (5.164)$$

For the BCS – and finally SCS – motion, the wind velocity in the BCS is modeled as a bias vector δ_W, yielding:

$$^{SCS}\dot{p} \;=\; {}^{SCS}BCS \cdot \left(^{WCS}v + \delta_W \right) \qquad (5.165)$$

Please note that δ_W can also model vertical wind, which might actually be beneficial in mountainous terrain. Otherwise, δ_W^z also compensates for non-neutral lift trim of the airship, by suggesting a "virtual" vertical airflow. Unfortunately, the latter "virtual" airflow would need to depend on v^*, but anyway, there is just no simple way to reliably distinguish trim effects from real vertical airflow.

This already completes the simplified "roller-coaster kinematics" model underlying the control of the airship.

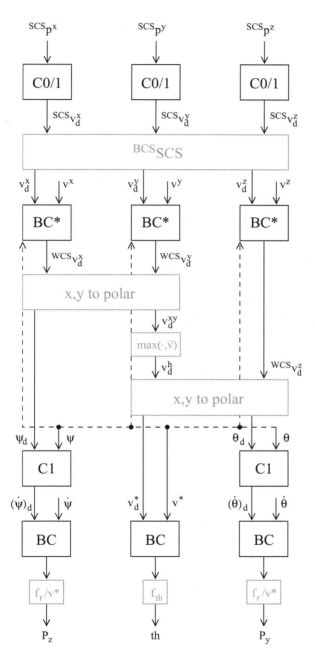

Figure 5.28: Proposed control architecture for an airship UAV, assuming "roller-coaster kinematics".

5.6.2   Control Architecture

Figure 5.28 shows the control architecture "inverting" the airship model from section 5.6.1. The top part consists of three *C0/1* integrator stage mode-switching controllers governing (5.165) by requesting a desired velocity $^{SCS}v_d$, which is instantly transformed into the BCS (v_d). In the case of *C0* "traveling" mode along SCS-x, the control stage will output a fixed desired velocity $^{SCS}v_d^x = \bar{v}_d$, independent of the state.

The desired BCS velocity is now subject to "smart" bias compensation according to figure 5.9. The three integrators will converge to $-\delta_W$, as will be discussed in more detail below. Thus, the bias compensation stage's output is $^{WCS}v_d$, the desired airspeed vector, as obtained from inverting (5.165).

Now, the basic task at hand is to convert $^{WCS}v_d$ into polar coordinates, inverting (5.164). But at this point, it is necessary to take care that the magnitude v^* of the actual airspeed cannot get too low, for two reasons:

1. The model introduced in section 5.6.1 relies on some lower bound on v^*, as explained in detail above.

2. Even considering any model, an airship with a constant direction of its thrust vector becomes effectively immaneuverable with v^* too low. At $v^* = 0$, it cannot exert any controlled force in both VCS-y and VCS-z directions.

Therefore, the transform to polar coordinates observes that the magnitude v_d^h of the BCS-horizontal projection of the desired airspeed is kept above some bound $\tilde{v} > 0$:

$$\psi_d = \operatorname{atan2}\left(^{WCS}v_d^y, \; ^{WCS}v_d^x\right) \tag{5.166}$$

$$v_d^{xy} = \sqrt{\left(^{WCS}v_d^x\right)^2 + \left(^{WCS}v_d^y\right)^2} \tag{5.167}$$

$$v_d^h = \max\left\{v_d^{xy}, \tilde{v}\right\} \tag{5.168}$$

$$v_d^* = \sqrt{\left(v_d^h\right)^2 + \left(^{WCS}v_d^z\right)^2} \tag{5.169}$$

$$\theta_d = \operatorname{atan2}\left(^{WCS}v_d^z, \; v_d^h\right) \tag{5.170}$$

This special artifice ensures that the controller never "tries" to fly too slow, while both the horizontal direction (yaw angle) of the desired velocity and its vertical component are not affected. It should be noted that the step (5.168) *violates* the controller design procedure, because it is not reflected in the plant model (and not invertible, by the way). The observable result of this solution will be examined later. If and only if the maximum operation does not apply, i.e. $v_d^h = v_d^{xy}$, above equations *exactly* invert (5.164) as usually required.

The remaining calculations of the controller first apply *C1* control to the yaw and pitch angles, yielding desired change rates as modeled by (5.161,5.162). A simple bias compensation stage is added to each of the rudder pitch controls in order to get rid of possible calibration issues.

---

[6]Of course, this violates (5.158), which demands $\alpha = 0$ in the equilibrium case. But the "violation" here is only *transient*, for the derivation of (5.161).

Finally, the inverse of the functional transform from (5.161,5.162) needs to be applied to yield the control inputs P_z, P_y. For throttle control, there is no integrator stage to negotiate. Therefore, only bias compensation applies, followed by the inverted proportional transform from (5.160). For bias compensation, the actual airspeed v^* must be provided by an airspeed sensor.

This already finishes the presentation of the control architecture, but leaves two details to discuss: The functioning of the "smart" integrators for wind estimation, and the consequence of v^*-thresholding by (5.168).

Wind Estimation

"Smart" bias compensation according to figure 5.9 cannot be directly applied here, because the thresholding function (5.168) is not invertible. Therefore, the "upper left" inputs to the integrators as depicted in said figure need to be fed with $^{WCS}v_d$ directly. Hence, the controller's robustness against rounding errors is not as good as discussed in section 5.3.4. However, closed-loop characteristics degrade to input-output stability, anyway, with v^*-thresholding active. Thus, the optimal solution in this special case is to use the "smart" bias compensator strictly according to figure 5.9 whenever the threshold does *not* apply, and use the direct $^{WCS}v_d$ input whenever it *does*.

Consequence of v^*-Thresholding

Considering a straight-line approach to a target point p_+ in calm air, v^*-thresholding will prevent the vehicle from decelerating and cause it to overshoot the target. At that moment, the yaw angle error jumps to roughly 180°, which will initiate a turn (in random direction) and a new approach with $v_d^* = \tilde{v}$ as the desired airspeed. This procedure would similarly repeat over and over again, resulting in a trajectories consisting of "figure-eights" with their center at p_+ and varying rotational directions.

However, this extra examination will demonstrate that the system behavior in such cases can be changed into the airship's flying a horizontal circle around p_+. This circle will be referred to as *standby circle* in the remainder of this examination. The standby circle will appear provided that the $\dot{\psi}_d$ summand for propagating the changes of the desired value is removed from the control equation for ψ, i.e. leaving only

$$(\dot{\psi})_d \;=\; k \cdot (\psi_d - \psi) \tag{5.171}$$

instead of the usual

$$(\dot{\psi})_d \;=\; k \cdot (\psi_d - \psi) \;+\; \dot{\psi}_d. \tag{5.172}$$

Figure 5.29 introduces a state description relevant to the standby circle: r is the horizontally projected radius vector $p - p_+$ from the target point to the vehicle's location. v is the current horizontal airspeed vector of the vehicle; it is $\|v\| = \tilde{v}$, for the effect of thresholding is to be examined here. Finally, α is the angle between v and $-r$, negative in the depicted situation.

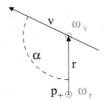

Figure 5.29: Derivation of standby-circle around target point for airship UAV.

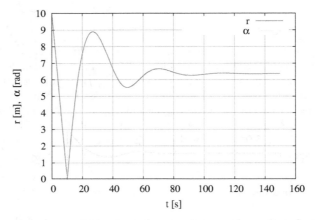

Figure 5.30: Simulated convergence process to standby circle ($k = 0.1\ \mathrm{s}^{-1}$, $\tilde{v} = 1\ \mathrm{m/s}$).

The system defined in this way can be described, in fully scalar notation, through:

$$\dot{r} = -\tilde{v} \cdot \cos\alpha \tag{5.173}$$
$$\dot{\alpha} = \omega_v - \omega_r \tag{5.174}$$

Here, ω_v is the turning rate of the velocity vector v, while ω_r is the turning rate of the radius vector r. Also considering the model and control for ^{WCS}v and ψ from above, one obtains

$$\omega_v = \dot{\psi} = k \cdot (\psi_d - \psi) = -k \cdot \alpha \tag{5.175}$$
$$\omega_r = -\frac{\tilde{v}}{r} \cdot \sin\alpha \tag{5.176}$$

and finally:

$$\dot{\alpha} = \frac{\tilde{v}}{r} \cdot \sin\alpha - k \cdot \alpha \tag{5.177}$$

The system (5.173,5.177) is *not* suitable to describe trajectories through the target point ($r = 0$), due to r in the denominator in (5.177). Nevertheless, convergence to the standby circle can be examined. A formal stability proof is out of the scope of this examination, but simulation results make clear that the standby circle is a stable attractor, especially after a straight approach to p_+. Figure 5.30 shows a simulation run starting at $r = 10$ m, $\alpha = 0.01$ with $k = 0.1$ s^{-1} and $\tilde{v} = 1$ m/s. While the straight line toward p_+ is a stable attractor during the first approach, convergence to the standby circle follows the turn after passing p_+, in a distinctly damped oscillation. From (5.173,5.177) it is clear that equilibrium states ($\dot{r} = \dot{\alpha} = 0$) are given by:

$$\alpha = \pm\frac{\pi}{2} \tag{5.178}$$

$$r = \frac{2\tilde{v}}{k\pi} \tag{5.179}$$

As the standby circle is an attractive system behavior for "hovering" in too little wind and it requires to disable the ψ_d term in P_z-control, a recommendable solution for an implementation of the airship controller would be to disable the ψ_d term if and only if the v^*-threshold rule is in effect. This solution has the additional benefit that too high $(\psi)_d$ signals – which can only happen with small v_d^{xy} – are automatically masked from control as a side effect.

5.7 Example Controller: Fixed-Wing UAV

This section presents a flight controller design for a fixed-wing UAV. Analog to the previous section, the emphasis is again on the design procedure, not on deriving the plant model nor on the experimental validation of the resulting controller. Nevertheless, the underlying model used here is less idealized than the one used for the airship controller. It turns out that there are surprisingly strong similarities between the airship controller (figure 5.28) and this one for fixed-wing UAV control.

5.7.1 System Model

The UAV considered here exhibits the following control inputs (see section 1.3.2):

- *th*, the engine control signal ("throttle"),

- P_x, the aileron pitch setting,

- P_y, the pitch setting of the elevator, and

- P_z, the pitch setting of the yaw rudder.

For the plant model, the effect of the yaw rudder setting is *not* considered. Here, too, it shall be assumed that the heading and WCS-velocity vectors are always parallel, i.e. (5.158) holds. Certainly, for a fixed-wing plane this applies with much less error than for an airship. However, the yaw rudder will be subject to open-loop control in a way to minimize *slip*, i.e. to keep (5.158) "as correct as possible".

The state description consists of the following components and is, analog to the MARVIN and airship models, parameterized by the current course segment coordinate system SCS:

$$S_{SCS} = \begin{pmatrix} {}^{SCS}p \\ \psi \\ \theta \\ \phi \end{pmatrix} \in \mathbb{R}^6 \tag{5.180}$$

The components of S_{SCS} denote:

- ${}^{SCS}p = {}^{SCS}BCS \cdot (p - p_+) \in \mathbb{R}^3$ is the plane's SCS position (i.e. position error), just as defined in (5.107) for MARVIN.

- $\psi, \theta, \phi \in (-\pi; \pi]$ are the "standard" yaw, pitch, and roll Euler angles of the plane, corresponding to (5.108).

The remainder of this subsection presents the model equations describing the behavior of S_{SCS}. For throttle th and equilibrium speed v^* as in the previous section, (5.160) from the airship model shall be extended by a force component $\Delta F \sim \sin \theta$ dependent on the current pitch angle θ of the plane, yielding:

$$(v^*)^2 = (1/f_{th})^2 \cdot th^2 + f_\theta^2 \cdot \sin \theta \tag{5.181}$$
$$v^* = \sqrt{(1/f_{th})^2 \cdot th^2 + f_\theta^2 \cdot \sin \theta} \tag{5.182}$$

Here, f_θ is a constant technically expressing the equilibrium "free-fall" velocity of the plane.

There is some additional significant interrelation between v^*, the attack angle of the vehicle's airfoils required for equilibrium lift, and the airfoils' aerodynamic drag, which is difficult to model (see e.g. [38]). Therefore, these effects will be left to a bias compensation stage within the controller.

The effects of the aileron and elevator rudders are assumed to be proportional to the plane's corresponding rotation rates, following the same consideration as underlying (5.161,5.162). But in the case of $\dot{\theta}$, the effect of gravitation needs to be separated from the effect of the elevator. This can be done by introducing the lift acceleration a_L, which is the virtual acceleration as measured by an acceleration sensor in the VCS-z axis. Then:

$$\dot{\phi} = {}^{VCS}\omega^x = (v^*/f_x) \cdot P_x \tag{5.183}$$
$$a_L = ((v^*)^2/f_y) \cdot P_y \tag{5.184}$$

Here, a_L would result in a change of θ, but this change is also influenced by the plane's weight, which just neutralizes a_L in the equilibrium case. Formally:

$$\dot{\theta} = \frac{a_L}{v^*} \cdot \cos \phi + \frac{g}{v^*} \cdot \cos \theta \tag{5.185}$$

Finally, the change rate of the yaw angle ψ needs to be modeled. The BCS-horizontal acceleration a^{xy} affecting ψ solely results from a_L. Considering $a^{xy} = -a_L \sin \phi = \dot{\psi} v^{xy}$ and $v^{xy} = v^* \cos \theta$, one obtains:

$$\dot{\psi} = -\frac{a_L}{v^*} \cdot \frac{\sin \phi}{\cos \theta} \tag{5.186}$$

The inversion of (5.185,5.186) for a_L and ϕ will be demonstrated below during the presentation of the control architecture.

With ψ, θ, v^* given, the remaining state variables develop as in the airship model, according to (5.164) and (5.165). The wind is again regarded as defining a separate coordinate system through a velocity bias δ_W.

5.7.2 Control Architecture

Figure 5.31 shows the control architecture "inverting" the plane model from section 5.7.1. The top part equals the airship controller, with the exception that $^{SCS}v_d^x$ is constant instead of calculated by a mode-switching control equation. This indicates that the plane controller as presented here does not use an actual target *position*, but only a current course segment to be followed indefinitely[7].

As v^*-thresholding is not applied in this controller, $^{WCS}v_d$ is directly transformed into polar coordinates, inverting (5.164) through:

$$\psi_d = \text{atan2} \left(^{WCS}v_d^y, \ ^{WCS}v_d^x \right) \tag{5.187}$$

$$v_d^* = \sqrt{\left(^{WCS}v_d^x \right)^2 + \left(^{WCS}v_d^y \right)^2 + \left(^{WCS}v_d^z \right)^2} \tag{5.188}$$

$$\theta_d = \text{atan2} \left(^{WCS}v_d^z, \ \sqrt{\left(^{WCS}v_d^x \right)^2 + \left(^{WCS}v_d^y \right)^2} \right) \tag{5.189}$$

Two standard *C1* integrator stage controllers yield the desired change rates $(\dot{\psi})_d$ and $(\dot{\theta})_d$. Now, the inversion of (5.185,5.186) is required, indicated by *to VCS* in the control architecture. By resolving these two equations for $\sin \phi, \cos \phi$ and using atan2, the term a_L drops out nicely, leaving:

$$\phi_d = \text{atan2} \left((\dot{\psi})_d v^* \cos \theta, \ g \cos \theta - v^* (\dot{\theta})_d \right) \tag{5.190}$$

$$a_{L,d} = \frac{g \cos \theta - v^* (\dot{\theta})_d}{\cos \phi_d} \tag{5.191}$$

These equations are defined, at least, for desired $\dot{\theta}$ not exceeding the effect of gravity.

ϕ_d from above is then fed through another *C1* control to yield the desired roll rate $^{VCS}\omega_d^x$, which can be effected via the ailerons. Then, the three desired values v_d^*, $^{VCS}\omega_d^x$, and $a_{L,d}$ are subject to "simple" bias compensation according to figure 5.8, in order to eliminate remaining

---

[7]Switching to the next segment is not considered here, but could of course be performed by computing a "virtual" position control, for comparison only, and applying the rules from section 5.5.5.

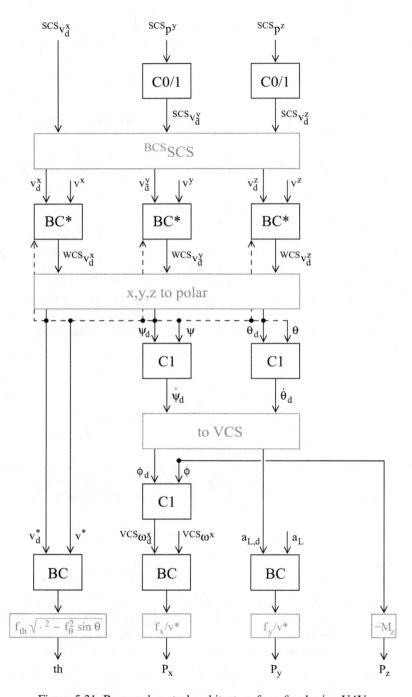

Figure 5.31: Proposed control architecture for a fixed-wing UAV.

model and calibration errors. For completing the inputs to this bias compensation stage, the actual values need to be measured by an airspeed sensor, a VCS-x rotation sensor, and a VCS-z acceleration sensor.

Finally, the following inverted transforms yield the actual control inputs:

$$th = f_{th} \cdot \sqrt{(v_d^*)^2 - f_\theta^2 \cdot \sin\theta} \qquad (5.192)$$
$$P_x = (f_x/v^*) \cdot {}^{VCS}\omega_d^x \qquad (5.193)$$
$$P_y = \left(f_y/(v^*)^2\right) \cdot a_{L,d} \qquad (5.194)$$

The last control input, the yaw rudder pitch P_z, is determined by heuristic feed-forward control aimed at minimizing the slip of the plane. Using a constant "mixing" coefficient M_z, it is:

$$P_z = -M_z \cdot \phi \qquad (5.195)$$

This completes the fixed-wing UAV controller.

5.8 Subsumption

In this section, an iterative way of constructing controllers for nonlinear systems of arbitrary order has been formally founded, examined, applied to three example systems from the UAV sector, and validated by means of the real-world performance of the MARVIN flight controller. Even without experimental validation, proposition 4 assures that the last two sample applications lead to GAS closed-loop behavior[8], given that the employed models are correct. Furthermore, the closed-loop convergence will be 100 % free of overshooting, no matter what gains k_j are chosen for integrator-stage control, provided there are no dead-times involved.

The three sample controllers have pointed out that the principal challenges in constructing a controller according to the proposed schema are:

1. to represent the plant through a model with invertible functional transforms, and

2. to insert the required bias compensation stages at places where the expected biases are as constant as possible.

Having already addressed helicopter, airship, and fixed-wing flight control, all control tasks to be met in the field of UAV design should be solvable via the proposed procedure.

---

[8]GIOS in the presence of variable biases and/or desired-velocity-thresholding as applied in the control of the airship.

Chapter 6

Sensors

This chapter deals with the need for and selection of sensors to be used on-board a UAV. Basically, these sensor serve (at least) one of these two different purposes:

1. to determine the position and attitude of the UAV in order to provide the required inputs to the flight controller, or

2. to measure data that is relevant to the mission that the UAV is fulfilling.

Section 1.1.2 in the introduction has already listed possible sensor types for the second item, the acquisition of mission-relevant data. In this chapter, possible sensors will be examined in more detail from a technical point of view. Clearly, some sensors may be involved in both of the above purposes. For instance, attitude and position data may also be required in order to properly map images obtained for the mission to real-world positions, or a magnetic field sensor might be involved in orientation measurement as well as in the detection of metallic objects.

This chapter starts, as usual, with a summary of requirements to on-board sensor equipment (section 6.1). Then, section 6.2 deals with specific sensors *primarily* used for flight control purposes. Section 6.3 discusses the mathematical task of fusing measurements from "virtually" redundant sensor groups to obtain optimal quality – and possibly fault-tolerant – position and attitude data. Section 6.4 finally addresses image sensors, being the most frequently desired type of sensors for mission data acquisition – other kinds of mission-specific sensors are not covered in this book, because this cannot be done with sufficient generality. The chapter finishes with a short subsumption of the findings.

6.1 Requirements

This section lists requirements to be typically met by on-board sensors of UAVs. The requirements discussed here are mainly focussed on the selection of suitable sensor devices. The sufficiency of certain sensors for a particular mission task is not covered, for possible missions types are innumerable and the suitability of a sensor can only be decided based upon a detailed task specification. The requirements review is divided into general requirements and requirements specific to attitude measurement.

6.1.1 General Requirements

General requirements to UAV on-board sensors are roughly analog to the requirements on all on-board systems for UAVs. Specifically, they are:

- Sensors must obey strict *limitations* regarding weight, size, power consumption, and (if possible) cost.

- Sensors need to be equipped with an *interface* suitable for integration into the on-board information exchange. This issue has already been discussed in some detail in section 3.1.2.

- It should be assured that the *bandwidth* of the data stream acquired by a particular sensor does not exceed the capabilities of on-board processing and/or wireless transmission to the ground station. In other words, it is a waste of resources to employ on-board sensors the performance of which cannot be properly exploited. While this fact seems trivial, it has not always been obeyed by UAV designers. This issue will be discussed below in the context of on-board cameras.

6.1.2 Requirements of Attitude Measurement

For UAV flight control, it is evident that 3 DOF (degrees of freedom) position data and 1 DOF orientation data constitute the absolute minimum sensor data input to flight control. This results from the fact that 3 DOF position control will of course be expected from any autonomous UAV, and yaw angle measurement will be required to exert propulsion forces in the correct direction. Yaw is special here because no external effect is capable of passively stabilizing the yaw angle, while pitch and roll may be passively stabilized due to gravity. This is the case, for example, in the airship model presented in section 5.6.1. Passive stabilization, though, can only apply to UAVs that exert lift forces not anchored in the vehicle coordinate system (VCS as defined in section 5.5.1). Lift forces exerted through rotors and wings, on the other hand, are fixed in the VCS and therefore incapable of stabilizing any orientation angle.

Special cases of UAVs are imaginable that allow orientation estimation from velocity data together with the plant model. This particularly applies to fixed-wing UAVs, which are aerodynamically forced to roughly align their orientation with their traveling direction. Then, model-based orientation estimation from the velocity vector is possible without any dedicated orientation sensor. In the context of this chapter, model-based estimation techniques of this kind shall be regarded as orientation *sensors* as well.

In the most general case, orientation measurement in full 3 DOF is required. The following theoretical consideration is fundamental to the design or selection of suitable sensors or sensor groups for this purpose:

- First, *rotation rate* sensors are theoretically sufficient for all purposes of orientation measurement. If the orientation variables in the state vector are correctly initialized at the beginning, orientation can be permanently updated by *integrating* the measured orientation rates about the VCS axes. However, orientation sensors according to this approach are subject to *drift error*, i.e. a small sensor bias in the rotation rate measurement might integrate to arbitrary large orientation errors over time, rendering the sensor subsystem itself unstable. Therefore, one must either use very accurate sensors and integration algorithms and limit the operation time of the UAV in order to bound the drift error, or otherwise implement some means of drift compensation.

- *Acceleration sensors* are often used – or at least regarded – as a potential way of obtaining drift-free measurements of the pitch and roll angles, since the measured acceleration points "down". This is, however, an oversimplification, as will be explained now. According to Newton's law, the exerted forces F_e from rotors, airfoil, and possibly static lift on the one hand, and the weight $m \cdot g$ on the other, combine into a resulting force F that leads to an acceleration a of a vehicle, in vector notation, with:

$$F = m \cdot a = m \cdot g + F_e \qquad (6.1)$$

In this equation, gravity g itself *cannot* be measured by any pseudo-force type acceleration sensor, because it affects all parts of the sensor in the same way, which is fully equivalent to *not at all*. Acceleration sensors mounted on the vehicle will measure $-F_e/m$ from above, because this is exactly the external influence that is directly coupled only to the "frame" of the sensor, while the coupling force between this frame and the sensor's "reference" mass is measured. Hence, the sensor measurement a_S actually is

$$a_S = g - a, \qquad (6.2)$$

converted into the VCS. So it is clear that the orientation information contained in the ^{VCS}g summand above can *only* be determined after correcting the measurement by the motion acceleration ^{VCS}a.

Another way of looking into the same fact is to imagine a UAV that exerts forces solely along VCS-z and moves through tilting, just like an idealized helicopter. When it tilts, the acceleration measurement a_S will still be parallel to VCS-z, so the orientation change is fully latent to acceleration sensors. To be fully precise, the tilting will effect in BCS-horizontal motion acceleration, which will finally create air resistance that adds to F_e and *might* make a_S converge to g again. But any orientation sensor relying on this aerodynamic equilibrium would obviously induce far too much lag to qualify for flight-control purposes.

On the other hand, if there is some upper-level position control in force that integrates position control errors in some way and adapts the desired orientation angles suitably, then this adaptation will compensate for *short-term* effects of orientation sensor drift, while (6.2) *can* be used for *long-term* sensor bias compensation. This is true because motion acceleration a averages to zero, $\bar{a} = 0$, due to position control, so $\bar{a}_S = g$[1]. This

---

[1]But please note that this averaging law strictly holds in the BCS only, since $^{VCS}g(t)$ is *not* constant.

also holds if only the orientation control stage is used, supporting a human pilot that commands the desired angles. In this case, the human pilot plays the role of an upper-level position controller.

Summing up the meaning of acceleration sensors to orientation measurement, it must be stated that they do *not* provide *any* orientation information unless a working position sensor is available as well.

- Even if acceleration sensors are properly used for rotation sensor drift compensation according to (6.2), they only provide 2-DOF orientation information – the yaw Euler angle ψ as defined in (5.108) cannot be obtained from the direction of ^{VCS}g. Hence, a drift error compensation reference for yaw is most suitably provided by *magnetic field sensors* measuring the magnetic field of the earth. Fortunately, this measurement can be performed directly, instantly, and without being subject to drift error. However, magnetic field sensors alone are *not* sufficient for drift compensation, because they, too, only yield 2-DOF attitude information – the magnitude of the measured field vector does not carry any attitude information.

Summing up above findings, rotation rate sensors are the primary source of orientation data but tend to require some means of drift error compensation. Drift error compensation is mathematically possible either through a combination of acceleration, magnetic field, and position sensors, the latter possibly being part of a working position control loop, or through a position (i.e. velocity) sensor in combination with a plant model.

Another, completely different approach to orientation measurement is through environment tracking, e.g. optical horizon detection. This is attractive in its simplicity, but limited to applications with predictable properties of the environment. Horizon detection, for example, could be fatally disturbed by unexpected obstructions through buildings or mountains that reach or exceed the vehicle's operating altitude.

Finally, it is important to consider *fail-safe procedures* if sensors with limited reliability are involved in providing input to flight control. Sensors like GPS and optical tracking certainly belong into this class, because they might fail at least temporarily due to external factors like radio interference, obstruction, or unsuitable environment properties. Then, fail safe procedures need to be available to generate suitable controller inputs from the remaining sensor signals to assure safe operation for the time of failure. An obvious example of such a fail-safe procedure is to use the integrated rotation rates only, because rotation rate sensors do not depend on external factors and their drift error may be tolerable *for a limited time*.

6.2   Position and Attitude Sensors

In this section, different individual sensor devices mainly suitable for position and attitude measurement are presented.

6.2.1 GPS

GPS, the satellite-based *global positioning system*, provides the most obvious means of position measurement for UAVs operating outdoors. The facts within the following GPS technology overview are taken from the current Standard Positioning Service Performance Standard (SPS) [16] and the book [110].

The Navstar Global Positioning System, operated by the US Department of Defense, consists of nominally 24 satellites orbiting approximately 20,000 km above the earth's surface, plus a varying number of "spare" satellites. During the setup of GPS, the first satellite ("Block I" type) was launched in 1978, and the full 24 satellite constellation ("Block II, IIA, IIR" types) was completed in March 1994, rendering the system fully operational. All satellites transmit two carrier signals, called L1 at 1.57542 GHz and L2 at 1.2276 GHz, and provide two different positioning services, the *standard positioning service* (SPS) available to all – including civilian – users, while the *precise positioning service* (PPS) is reserved for military use. For SPS, a 50 bit/s navigation message is XORed to a 1.023 MHz *pseudo-random-noise* (PRN) code sequence, the *coarse/acquisition* (C/A) code. The resulting signal is modulated onto the L1 carrier to yield a spread-spectrum signal. The encrypted PPS signals are transmitted via both L1 and L2 carriers.

The satellite constellation constitutes the *space segment* of GPS. The ground-based *control segment* supervises the satellites and must occasionally upload new information for the generation of the navigation messages. For the former Block I satellites, such uploads were required every 3.5 days, while later satellites can operate autonomously for 180 days at least. The GPS receivers operated by agencies and private users, finally, constitute the *user segment*.

Before May 2000, SPS was intentionally deteriorated by random errors included in the signals, a procedure euphemistically termed *selective availability* (SA). Since then, SA has been disabled by presidential order. As a result, the average positioning error as resulting from the SPS signals' properties is specified to be within 13 m horizontally and 22 m vertically with 95 % confidence [16].

13 m position accuracy does not sound too promising for UAV flight control, especially if velocity needs to be calculated from subsequent position measurements. However, the following observation and existing augmentation methods do eliminate this concern:

- Above 13 m and 22 m errors refer to the absolute global positioning accuracy. Subsequent measurements are almost always subject to approximately the same absolute errors, so relative positioning accuracy as well as velocity measurement from secant slope are better by orders of magnitude. Velocities calculated from 1 Hz position updates usually exhibit noise levels in the order of magnitude of 0.1 m/s only (also depending on the receiver used, of course).

- By using a reference receiver at a fixed location in the vicinity of the moving receiver, both receivers are subject to roughly the same perturbations. This leads to the approach of *differential GPS* (DGPS), in which standardized reference signals are transmitted from the reference receiver to the moving receiver to correct its measurement for the

Figure 6.1: u-blox-Antaris-4-based GPS Module (Conrad CR4).

common perturbations. This leads to absolute global errors usually in the order of magnitude between 1 m and 3 m.

- DGPS can even be augmented by tracking carrier-phase differences between the reference and moving receivers in addition to the navigation messages' timestamps. This is possible for any receiver for both L1 and L2 carriers, because no demodulation nor signal decoding is required. With wavelengths in the order of magnitude of only 25 cm, this *carrier-phase DGPS* technology combining L1 and L2 reception is able to reduce even absolute errors to as low as 2 cm. While this absolute accuracy requires the position of the reference station to be known at least as well, the same accuracy can be easily obtained relative to the reference station, which is fully sufficient for most UAV application scenarios.

Depending on the accuracy requirements of flight control in the case of a particular UAV, reasonable GPS equipment options vary considerably. Today, a great variety of light, small, and cheap embeddable receivers are available, many of them primarily developed for street navigation applications. However, one major drawback is present in *almost* all of these devices: Their maximum rate of position updates is limited at 1 Hz. This *might* be sufficient for certain types of UAVs, especially fixed-wing vehicles, provided that the control gains are kept moderate. But a higher update rate will be desirable in any case.

Currently, there is mainly one source of low-cost sensors available that do not exhibit the 1-Hz-limitation, the Swiss-based manufacturer u-blox. Selecting their currently most attractive design for small UAV applications, the Antaris 4 chipset [135], pinpoints e.g. the *CR4* USB receiver [121] by Conrad, Germany, with the following set of features:

- 16-channel GPS receiver with integrated patch-antenna,

- position update rate up to 4 Hz,

- standard DGPS support (RTCM input, WAAS, EGNOS),

- 1 USB port (with cable), 1 RS 232 port (on PCB only),

- operating voltage 3.7...6.5 V, USB-powered

- supply current 39 mA typically (at 1 Hz update), 70 mA peak

- size 32 mm × 32 mm × 9.5 mm,

- mass approx. 20 g,

- single-unit price 62 EUR (May 2007).

Aside from carrier-phase DGPS, this device should be able to fulfill most small UAV position sensing requirements at really marginal resource employment. Figure 6.1 visualizes the module.

Unfortunately, advancing to carrier-phase DGPS results in a disproportionate cost increase, presumably due to the very limited number of applications actually requiring real-time accuracy in the range of centimeters. Manufactures involved in this technology include Trimble and NovAtel, both located in Canada. The NovAtel *OEMV-2* receiver card [75] shall serve as a current example of a two-frequency carrier-phase DGPS receiver, providing the following features:

- 14-channel dual-frequency GPS (or 12-channel dual-frequency GLONASS) receiver,

- max. position accuracy (L1/L2 carrier-phase DGPS) 1 cm CEP,

- max. update rate 20 Hz,

- 2 RS 232, 1 CAN bus, 1 USB ports,

- power consumption 1.6 W typically,

- size 60 mm × 100 mm × 13 mm,

- mass 56 g,

- estimated[2] price > 10,000 EUR.

It must be noted that a second receiver card with the same features is required to operate the reference station on the ground, and each of the receivers is to be connected to a suitable antenna in addition.

All in all, L1/L2 carrier-phase DGPS equipment may easily contribute more than 50 % of the UAV system's hardware cost. Hence, confinement to "conventional" DGPS should be the primary goal here.

---

[2]Price lists are available on request only. The predecessor OEM4 board with comparable features cost about 10,000 EUR in 2005.

Aside from GPS, the "competing" Russian GLONASS and the future European Galileo systems would or will, respectively, provide roughly the same capabilities. Therefore, the key performance features and their affordability, as mentioned above for GPS, actually apply as general guidelines regarding satellite-based position sensors.

6.2.2   Rotation Rate

Rotation rate sensors, sometimes referred to as "gyroscope sensors", are the most relevant source of attitude data, as explained above in the requirements section 6.1. The term "gyroscope" stems from the oldest technical implementation as an actually rotating gyroscope, with the precession torque resulting from rotation about an axis perpendicular to the gyro's main rotation axis being measured.

Today, rotation rate sensor do not need to show a rotating gyroscope any more. Instead, the following technical implementations are possible:

- *Laser gyroscopes* send two beams of coherent light through a ring (*ring laser gyroscope*) or one or two coils of optical fiber (*fiber optic gyroscope*), in opposite directions. Rotation around the axis of the ring or coil shortens or lengthens the way traveled by the light, and subsequent superposition of the two beams exhibits a corresponding phase difference. This class of rotation sensors does provide the best accuracy by far, with drift errors possible in the order of magnitude of only a few degrees per hour. However, their size and cost render laser gyroscopes mainly unattractive for UAV applications.

- *Vibrating structure gyroscopes* excite an oscillation of a probe mass. Any rotation about an axis perpendicular to the direction of oscillation causes a Coriolis force, which in turn induces an oscillation perpendicular to both the excited oscillation and the rotation. This induced oscillation is measured by capacitive, piezoelectric, or electromagnetic means. This class of rotation sensors is also used in "gyro" modules for tail rotor control in remotely piloted model helicopters. In quantities, these sensors are available at prices between 10 EUR and 30 EUR. However, their accuracy is limited so that integration drift errors in the order of magnitude of one degree per second may occur.

Thus, vibrating structure gyroscopes are the method of choice for UAV applications. Here, *piezoelectric* and silicon-based *microelectromechanical systems* (MEMS) implementations can be distinguished.

Piezoelectric sensors may work as depicted in figure 6.2: A rod made of piezoelectric crystal carries six printed electrodes (hatched). A sinusoid alternating exciting voltage U_e is applied to the leftmost electrode, causing the left part of the rod to contract and expand corresponding to U_e, thus exciting a torsional oscillation of the rod in the depicted x-direction. When rotated about z, Coriolis forces induce an oscillation component in y-direction, which can be measured through the voltage U_ω generated in the crystal. For the generation of the sensor output, signal components with other (especially lower) frequencies than that of U_e must be removed from the U_ω signal; these may result from linear acceleration along y.

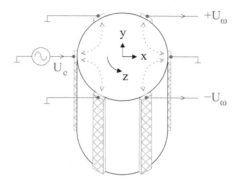

Figure 6.2: Construction scheme of vibrating structure gyroscope with piezoelectric rod.

This construction scheme is used by NEC/Tokin, for example in their current rotation rate sensor CG-L53, which shows the following significant features [33]:

- single-axis rotation rate sensor with analog output,

- measurement range $\pm 90\,°/s$,

- oscillating rod size 0.82 mm × 9 mm,

- supply voltage 3.0 V,

- supply current max. 4 mA,

- sensitivity 0.66 mV/($°/s$),

- package size 6 mm × 10 mm × 2.5 mm.

The first MEMS implementation of a rotation rate sensor has been reported in 2004 by Analog Devices [54]. Here, the probe mass is micromachined from polysilicon, tethered in a moving frame, and excited through an electrical field. The frame can itself move in a direction perpendicular to the resonance motion of the probe mass, the frame's displacement being detected capacitively by surrounding sense fingers. As this displacement is of sub-atomic scale, the post-processing electronics must be placed on the same substrate as the MEMS device and analyze the output signal from the detection fingers by correlation with the excitation signal.

The ADXRS300, one of the products using this technology, is characterized by the following features [14]:

- single-axis rotation rate sensor with analog output,

- measurement range $\pm 300\,°/s$,

- supply voltage 4.75...5.25 V,

- supply current max. 8 mA,

- sensitivity 5 mV/(°/s),

- package size 7 mm × 7 mm × 3.65 mm,

- unit price 30 USD (@ 1000 units).

6.2.3 Acceleration

Acceleration sensors are simpler to design than rotation sensors. They must also detect the deflection of a probe mass, but this probe mass does not oscillate, and the deflection tends to be much stronger than the one caused by the Coriolis force in vibrating structure gyroscopes. Therefore, MEMS implementations have been available for quite some time and constitute an obvious choice. The deflection is usually sensed through capacitive changes.

Analog Devices is also a reasonable source of MEMS accelerometers. As an example, the ADXL322 shall be considered here [15]:

- 2-axis acceleration sensor with analog output,

- measurement range ±2 g,

- supply voltage 2.4...6.0 V,

- supply current max. 0.5 mA,

- sensitivity 420 mV/g,

- package size 4 mm × 4 mm × 1.45 mm,

- unit price 3.75 USD (@ 1000 units).

For interfacing without the need for analog-to-digital conversion, sensors with PWM output are also available. The ADXL213 is an example, otherwise roughly comparable to the ADXL322.

6.2.4 Magnetic Field

As outlined in section 6.1.2 above, the long-term on-line calibration of attitude angles usually requires a magnetic field sensor, i.e. a compass. While a 2-DOF compass is theoretically sufficient to determine the yaw angle with pitch and roll angles known, e.g. from motion-corrected accelerometer readings, it is generally preferable to have a 3-DOF magnetic field sensor, for better numerical handling and to decouple accelerometer and magnetometer errors as far as possible.

A very affordable and fully sufficient class of magnetic field sensors are *fluxgate magnetometers*. The physical approach used here is a coil inducing an alternating magnetic field in a

Figure 6.3: "Home-made" 3-DOF compass using one FGM-2 sensor and one FGM-1 sensor by SCL.

ferromagnetic core, crossing the saturation level (or any other non-linearity of magnetic induction) in both directions of magnetization. The occurrence of this non-linearity is shifted by an external field, and the shift can be detected using a second coil for sensing the induced magnetic field's change rate [60]. One product well suited for UAV applications are the FGM-1 and FGM-2 sensors by Speake & Co. (SCL) [27]. The FGM-2 is a two-axes-sensor logically combining two single-axis FGM-1 sensors. The sensors' basic features consist of:

- PWM output with period proportional to field strength,

- measurement range $\pm 50 \ \mu$T,

- output frequency 50...120 kHz,

- supply voltage 4.5...7.0 V,

- supply current 12 mA,

- package size (FGM-1) 35 mm $\times$ 8 mm $\times$ 8 mm,

- single-unit (or per-axis) price 25 EUR.

Figure 6.3 depicts a "home-made" 3-DOF compass unit consisting of one FGM-2 and one FGM-1 and some circuitry, including dividers to reduce the PWM output frequency. This unit has been successfully used for attitude measurement aboard an older design stage of the MARVIN UAV.

Alternatively, magneto-resistive (MR) semiconductor devices with analog output may be used, such as e.g. the HMC1001 by Honeywell. They are available in 2- and 3-axis combined variants as well. One substantial disadvantage of this series of sensors is that for optimum performance, a periodic short high-current "reset" pulse is required, e.g. 3 A for 2 μs every 200 μs, realigning the magnetic domains in the MR sensor [111]. Some of the remaining features of this sensor are:

- measurement range ± 200 μT,

- supply voltage 5...12 V,

- supply current (depending on reset) $\approx$ 50 mA,

- sensitivity (@ 5 V supply, 3 A reset) 16 mV/gauss,

- package size 11 mm $\times$ 4 mm $\times$ 2 mm,

- single-unit price 20 USD.

Instead of individual sensor elements, integrated "intelligent" 3-DOF compass sensors with RS 232 or similar interface are available as well, for example from the manufacturers Honeywell or Crossbow. While devices like these are often priced in the area of 1000...2000 EUR, they may be advantageous because of calibration issues: Magnetic field sensors are much more difficult to calibrate in a laboratory environment than acceleration or rotation rate sensors, because known reference fields are not easily obtained. "Intelligent" sensors are usually factory-calibrated with regard to sensor gain, bias, and non-linearity.

Nevertheless, even an ideal magnetic field sensor is adversely affected by ferromagnetic material in the structure of the UAV. Therefore, *calibration procedures* must usually be implemented unless the vicinity of the sensor is ideally non-magnetic. There are two different kinds of magnetic perturbation to be considered:

1. *Hard-iron perturbation* results from a magnetic field constant in the VCS, stemming from magnetized material. This kind of perturbation is relatively easy to compensate because the perturbation field additively overlays the earth's magnetic field.

   In-field calibration of an on-board magnetometer can be performed solely based on the magnetic field of the earth, assuming that the direction of the field vector but not the field intensity is known. For this purpose, the sensor output $B_S \in \mathbb{R}^3$ is assumed to relate to the true field vector $B \in \mathbb{R}^3$ via

   $$B \;=\; C \cdot B_S + D, \tag{6.3}$$

   with $C \in \mathbb{R}^{3 \times 3}$ and $D \in \mathbb{R}^3$ denoting calibration parameters. (6.3) may account for any combination of per-axis biases and per-axis gains of the three sensors, and for any orientation and misalignment of their axes (of course, they should be "roughly" perpendicular for numerical reasons). As C and D introduce 12 DOF altogether, 4 measurements $B_{j,S}$ with known field B_j are the minimum to determine C and D. Considering that attitude measurement requires direction information only, The actual "known" vectors B_j may be scaled to some arbitrary (but constant) length, which will then be incorporated in the resulting (C,D) pair. Hence, a mechanical compass for determining magnetic north plus the knowledge of the local inclination angle are fully sufficient for calibration according to (6.3).

Or course, the sample measurements for calibration need to vary in *all* B components. For better robustness of the calibration process, it is furthermore desirable to perform more than 4 sample measurements $B_{j,S}$. C and D can then be determined using a standard least-square-error approach, stating the error as

$$E \;=\; \frac{1}{2}\sum_{i=1}^{N}(C \cdot B_{i,S} + D - B_i)^2 \tag{6.4}$$

and solving

$$\nabla E \;=\; 0, \tag{6.5}$$

which constitutes a system of 12 linear equations in 12 unknowns and has a unique solution whenever the sample orientations where suitably chosen. For the MARVIN UAV, hard-iron compass calibration is performed with $N = 8$, using orientations facing magnetically north, west, south, and east, once with the rotor axis vertical and once with a roll angle of 90°, respectively.

2. *Soft-iron perturbation* results from ferromagnetic material in the UAV's structure that is variably magnetized depending on the current attitude. Soft-iron calibration is much more difficult than hard-iron calibration because the local magnetization is much more difficult to model. In navigation, this kind of perturbation is called *magnetic deviation* and traditionally solved through a *deviation table* mapping compass readings to true headings.

   For air vehicles, the deviation table should actually be two-dimensional, which makes both its determination and its application undesirably complex. However, in the case of the MARVIN UAV the use of a one-dimensional tabular correction has proven very suitable, so this approach shall be briefly explained here. The deviation table is recorded during a 360° turn of the vehicle about VCS-z in fully horizontal attitude. During this turn, the "true" heading angle cannot be obtained from the compass; instead, it is estimated from the integrated output of VCS-z rotation sensor. The recorded correction table models a function

$$V \;:\; \mathbb{R} \to \mathbb{R}^3 \tag{6.6}$$

mapping the "candidate" VCS compass angle β to a 3-dimensional offset of the field vector, so that the corrected reading B can be functionally obtained in the application phase by:

$$B \;=\; B_S + V\left(\mathrm{atan2}(B_S^y, B_S^x)\right) \tag{6.7}$$

Of course, this correction procedure is only approximate and restricted to "almost" horizontal attitude. Though for helicopter flight, this assumption is fulfilled with relatively little error. In the MARVIN case, (6.7) has proven to significantly improve compass accuracy.

If both hard-iron and soft-iron compensation are to be performed, then the recommended approach is to use first (6.3) in the measurement cascade to obtain the best-possible 3-DOF linear correction. Afterwards, (6.7) is applied to minimize remaining nonlinear deviations.

6.2.5   Sonar

Both in the case of autonomous landing and for obstacle avoidance, sonar is a useful sensor to measure the altitude above ground. It is certainly the easiest and cheapest approach for obtaining this kind of information. Its main disadvantage consists in its limited range of only a few meters usually. This renders sonar unsuitable for true flight-path planning applications.

A good recommendation concerning the selection of sensors is to rely on the technology originally developed by Polaroid for autofocussing their ancient instant cameras. While these cameras have mostly vanished, fortunately, their sonar sensors have become an appreciated state-of-art component in many robotic applications.

Today, sensor modules of this family are distributed by SensComp, Inc. The *Series 600 Smart Sensor* [124] is a disc-shaped ultrasonic transducer readily equipped with the necessary interfacing circuitry just on its rear side. Its features include:

- measurement range 0.15...11 m,

- operating frequency 50 kHz,

- beam angle $\pm 15°$,

- supply voltage 6...24 V,

- supply current 55 mA,

- transmit current 2 A,

- TTL-compatible trigger and echo signals,

- unit size 43 mm $\times$ 43 mm $\times$ 24 mm,

- mass 19 g,

- single-unit price 70 USD.

Figure 6.4 shows this sensor in its mounting hole in the bottom plate of the MARVIN UAV's control box. The actually usable range depends on the ground texture. Over grass, reliable measurements are possible up to 5 m at least.

Figure 6.4: Series 600 Smart Sensor by SensComp, mounted in the bottom plate of the MAR-VIN control box (mounting hole diameter 39 mm).

6.2.6 Integrated IMUs

While the sensors covered so far measure a single physical value each, there are integrated "intelligent" sensors available that determine position and/or attitude information from a couple of different values and their intelligent fusion. Data fusion must always be performed, as explained in section 6.1.2 above. It is a design choice how much of this happens within separate "intelligent" sensor subsystems, and how much of this is programmed in the primary on-board computer:

- The use of "intelligent" subsystems reduces software complexity and facilitates system integration.

- The use of the primary on-board computer will usually save component costs and may be formally superior because of the availability of *all* sensor data at one place.

A very common type of integrated sensor are *inertial measurement units* (IMUs). They usually combine rotation rate sensors, acceleration sensors, sometimes magnetic field sensors, and possibly even a GPS receiver to output processed attitude angles.

In former times, precise IMUs were usually constructed with *gyro-gimbaled* platforms in constant orientation carrying the (acceleration) sensors. This primarily helps to minimize numerical integration errors because no rotational transforms are required. In principle, the integration of acceleration measurements can be performed by capacitors, which completely overcomes the issue of discrete digitalization. But this type of IMU is not quite suitable in the small UAV sector due to size and weight issues. With processing technology advancing, these gimbaled IMUs are increasingly replaced by *strap-down IMUs*, with the sensors fixed "in the VCS" and processing equipment taking care of rotational transforms. In spite of still slightly lower performance, only this class of IMUs shall be considered here.

Most interestingly, many products of this family provide orientation data without access to position information. Regarding the consideration in 6.1.2, it is clear that they can only work in general as long as some superimposed position control is in force. This is somewhat surprising, particularly since this fact tends to be completely concealed by the manufacturers in their

Figure 6.5: MicroStrain 3DM-GX1 IMU.

manuals and data-sheets. Therefore, the designer should always consider reading only raw measurements from these sensors if possible (e.g. acceleration and angular rate) and have the on-board computer perform the gyro drift compensation formally correct according to (6.2).

A promising IMU product for small UAVs is the 3DM-GX1 by MicroStrain [74]. It combines accelerometers, rotation rate sensors, and magnetometers in a package of 90 mm × 64 mm × 25 mm weighing only 75 g (30 g without enclosure). However, when used on-board MARVIN, the accelerometer output exhibited an undesirably high sensitivity to vibration, which might result from a poorly designed hardware low-pass filter (yet this has not been fully analyzed). Anyway, the device is usable with MARVIN's flight control if mounted close to the vehicle's center-of-mass. The unit price is about 1500 EUR. See figure 6.5 for a picture.

Much bigger, and probably more carefully engineered, is the VG400 by Crossbow [37]. Its housing is a metal cube of approximately 9 cm edge length, the device weighs 640 g. Lacking magnetic field sensors, it only provides pitch and roll angles in its processed data output, but raw data output of acceleration and rotation rate is optionally possible as well. From the author's experience with this device, pitch and roll output angles are sufficiently good for helicopter UAV flight control with a closed position-control loop. Yet, the angle outputs tend to develop some asymptotic error during flight (a few degrees) that can successfully be avoided by centralized data fusion of the VG400's raw output, separate magnetic field, and position data. The unit price of this IMU is beyond 5000 EUR.

Both miniaturization and the necessity of centralized data fusion in mind, products providing only sensors and interfacing circuitry but no data processing are attractive. As an example, the *Crista OEM Sensor Head* by Cloud Cap Technology [26] provides 3 accelerometers and 3 rotation rate sensors (plus temperature sensors for gain and bias correction) together with 16-bit analog-to-digital converters. Host interfacing is via SPI bus and provides up to 10 kHz sampling rate. The module is sized 29 mm × 28 mm × 15 mm and weighs only 7 g. Price quotes are available on request only, expect a unit price around 700 EUR.

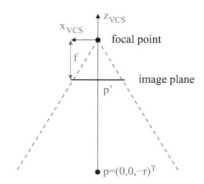

Figure 6.6: Optical flow model for projection beam $-z_{VCS}$.

6.2.7 Visual Navigation

When comparing automatic UAV flight control to the activity of a human pilot, it is obvious that one of the most relevant sources of information used by a human pilot has not yet been considered in the discussion of position and attitude sensors: visual perception of the environment. As already raised in the requirements outline, automated image processing demands considerable predictability of the environment properties in order to work with the required reliability. On the other hand, UAV operation indoors or in any area with GPS signal obstruction, given that no other external position sensor is provided, may depend on optical position and velocity sensors for navigation purposes.

Basically, two different kinds of optical navigation sensors are imaginable:

1. *Optical attitude sensors* may be implemented by tracking a feature at "infinite" distance, most probably the horizon. Due to this "infinity" assumption, the feature's location as recorded by an on-board image sensor only depends on the orientation of the vehicle, not on its position. This already introduces clear limitations to this approach, one of which consists in the possible existence of obstructing objects *too close* to the vehicle.

2. *Optical flow sensors* detect the 2-DOF motion of visual features within the image sensor's projection plane. Optical flow detection is a traditional task in computer vision [20], and many solutions have been proposed. Optical flow can be used to obtain information about the self-motion of the camera, but only in connection with other information sources, as will be explained here.

Figure 6.6 depicts the computation of optical flow v_I^x (as velocity in the image plane) for a camera looking in the direction $-z$ and a feature $p = (0,0,-r)^T$ exactly in this view beam (all coordinates are *VCS* here). The camera moves with velocity v^x. Then, it is:

$$v_I^x = -f \cdot \frac{v^x}{r} + f \cdot \omega^y \qquad (6.8)$$

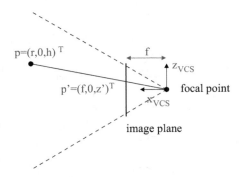

Figure 6.7: Optical flow model for camera view $+x_{VCS}$.

(6.8) shows that both rotation and linear motion of the vehicle contribute to the optical flow. $r \to \infty$ reflects the "infinite" distance case mentioned above, in which only the ω-summand is relevant. Otherwise, rotation rate sensors on board would permit to compensate for the ω-term. But even then, only the fraction v^x/r can be obtained from an optical flow measurement.

The same is true even for optical flow from a camera looking in the direction of motion. This is indicated in figure 6.7: Here, the camera faces in x-direction and tracks a feature at $p = (r, 0, h)^T$. With z' denoting the z-coordinate of the projected image and neglecting rotation (i.e. $\omega^y = 0$), one obtains:

$$z' = \frac{f \cdot h}{r} \tag{6.9}$$

$$v_I^z = -\frac{f \cdot h}{r^2} \cdot \dot{r}$$

$$= \frac{f \cdot h}{r^2} \cdot v^x$$

$$= z' \cdot \frac{v^x}{r} \tag{6.10}$$

Therefore, the forward-looking camera is also restricted to measuring v^x/r. In other words, the perceived velocity is always relative to the dimensions of the environment, which is not surprising with visual projections actually.

This is the reason why visual estimation of self-motion is not very attractive to UAV applications. It is possible, though, if an additional "range" sensor for determining r from (6.8,6.10) is provided. This may be a sonar, a laser range finder, or stereo imaging equipment (but note that the distance resolution of stereo imaging decays proportionally with the distance, so the resolution of velocity estimation from (6.8) together with stereo imaging would actually decay with r^2).

Summing this up, optical flow measurement may be beneficial mainly when flying very close to some fixed environment. Therefore, indoor operation is about the only application that both requires and permits optical position sensing.

The *HeliCommand Profi* by CAPTRON Electronic GmbH is an interesting "intelligent" control module for semi-autonomous helicopter UAV flight control [57]. It combines the following sensor groups in a device of 230 g:

- 3 rotation rate sensors,

- 2 optical flow detectors looking down,

- barometric altitude sensor,

- infrared distance sensor looking down.

The manufacturer states that the IR distance sensor operates up to 1.5 m above ground, and that position control based on the optical flow detectors works in altitudes up to 5 m (10 m "without wind"). These data are fully consistent with the findings listed above. Between 1.5 m and 10 m of altitude, the module most probably performs control of v/r in the sense of (6.8) instead of true velocity control, for the barometric sensor cannot determine the true altitude above ground during flight. This leads to slower motion in closer proximity to the ground, which in turn may even be considered desirable, if only for lack of an alternative. The set of sensors is sufficient, in any case, for autonomous landing. Velocity control is effectively performed in vehicle coordinates due to the lack of any world-anchored position reference. Hence, *HeliCommand Profi* is suitable as a human pilot support system, but not for truly autonomous position control. The module is prized around 3000 EUR, but stripped-down versions are available via modelsport supplier *Robbe* at about 300 EUR only.

6.3 Data Fusion for Attitude

In the requirement analysis above (section 6.1), it has been stated that attitude measurement typically

- involves rotation rate sensors, and

- must rely on other kinds of sensors for drift compensation purposes.

Tasks like this are generally referred to as *sensor fusion*. The most common approach is to use a *Kalman filter* for this purpose. A Kalman filter performs an optimal linear estimate of some observed state under the assumption of Gaussian noise with known parameters. Most important, it is a general approach that can usually be applied without substantial design effort. The well-known theoretical background, as set forth e.g. in [82], will not be addressed in this book.

The application of Kalman filters to different tasks of inertial and position data fusion has been thoroughly addressed in the literature. The issue of nonlinearity in the underlying state descriptions needs to be handled in some way. Traditionally, the measurement errors are used as the Kalman filter's state to "manually" linearize the system. More advanced methods directly apply an Extended Kalman Filter (EKF) to the position and orientation state variables, which is possible because an EKF uses a locally linearized model within the filter algorithm. More advanced approaches, e.g. [112], revert to direct filtering via simple linear Kalman filters through more sophisticated separation of dynamics and nonlinearity. This is desirable since an EKF is computationally more complex by an order of magnitude. Recently, effort has been made to apply sensor fusion already at the level of GPS pseudorange measurements instead of position data in order to be able to benefit from GPS information even if less than 4 satellites (minimum for position calculation) are currently visible, e.g. [56]. The details of these numerous approaches are without the scope of this book either, because choosing between them strongly depends on the requirements of a particular application.

Whenever a specialized Kalman filter is to be newly designed for a UAV application, the following word of warning may be appropriate: The generality of the Kalman approach, together with the general incorrectness of the Gaussian noise assumption, may lead to undesirable and unpredictable behavior nevertheless. The following statement, anonymously taken from a yet unpublished work, shall just serve as a hint to illustrate this kind of concern: *Large attitude errors can be a serious problem [...] after a GPS outage: Without any workarounds, a slow filter convergence, or even divergence can be the result.*

For this reason, and in order to demonstrate that the basic task of sensor fusion in attitude calculation is neither particularly complicated, critical, nor computationally expensive, the fusion algorithm used in the MARVIN UAV is presented and analyzed in more detail in this section.

Many reported sensor fusion algorithms use accelerometers also in order to provide position data in case of GPS outages. While this is formally possible of course, the necessary double integration of acceleration measurements with concurrently varying orientation renders these position data effectively unusable after very few seconds at best. This situation is particularly bad with low-cost and low-weight sensor equipment as typically used on-board small UAVs. Therefore, neither the MARVIN fusion algorithm nor the remainder of this book address any approach to position estimation from inertial sensor data. Instead, MARVIN's flight controller is designed to disable position control completely and revert to attitude control only when no GPS position data is available for more than one second. This is the best-suited failsafe procedure in such cases and will almost always avoid unsafe vehicle states for a longer time than position control on the basis of doubly integrated accelerations. Please not that according to section 6.1, the measured attitude will also degrade in case of GPS outage and become unsuitable after some time.

6.3.1   Fusion Algorithm Used with MARVIN

This section presents the sensor fusion algorithm used on-board MARVIN to calculate the orientation information from the readings of the angular rate, acceleration, magnetic field, and position sensors.

The presentation of the algorithms requires the introduction of another coordinate system beyond the ones defined in section 5.5.1, the *magnetic base coordinate system* (MCS). The MCS is the BCS rotated about z such that x points to local magnetic instead of true (geographic) north. As orthonormal matrices,

$$^{BCS}MCS = \text{Rot}_z(-D) \tag{6.11}$$

with D stating the angle of *magnetic declination* in the clockwise-is-positive notion being common for this purpose.

State Description

The state description S_a used for the attitude filter

$$S_a = \begin{pmatrix} ^{VCS}g \\ ^{VCS}N \end{pmatrix} \in \mathbb{R}^6 \tag{6.12}$$

consists of the vectors $^{VCS}g \in \mathbb{R}^3$ and $^{VCS}N \in \mathbb{R}^3$ that are supposed to point to $-z_{MCS}$ (down) and $+x_{MCS}$ (horizontally north), respectively.

From these two state vectors, MCS base vectors can easily be calculated as

$$^{VCS}z_{MCS} = \frac{-^{VCS}g}{\|^{VCS}g\|} \tag{6.13}$$

$$^{VCS}y_{MCS} = \frac{^{VCS}N \times {}^{VCS}g}{\|^{VCS}N \times {}^{VCS}g\|} \tag{6.14}$$

$$^{VCS}x_{MCS} = {}^{VCS}y_{MCS} \times {}^{VCS}z_{MCS} \tag{6.15}$$

and converted into VCS orientation by:

$$\begin{aligned} ^{BCS}VCS &= {}^{BCS}MCS \cdot {}^{MCS}VCS \\ &= \text{Rot}_z(-D) \cdot {}^{VCS}MCS^{\mathsf{T}} \end{aligned} \tag{6.16}$$

Please note that (6.14) above does *not* require ^{VCS}N to be exactly perpendicular to ^{VCS}g – in fact, any vector with a magnetically northward BCS-horizontal projection is suitable. This includes the magnetic field vector itself and will be exploited later.

State Update

The state update $S_{a,k} \rightarrow S_{a,k+1}$ requires to incorporate two different tasks:

1. apply rotation according to the integrated rotation sensor output, and

2. incorporate drift compensation (i.e. on-line calibration) based on gravity and magnetic field vectors, observing (6.2).

The second issue shall be discussed first.

Assuming standard GPS receiver output, the motion acceleration a involved in (6.2) has to be calculated from secant slope of subsequent velocity measurements, which in turn are obtained

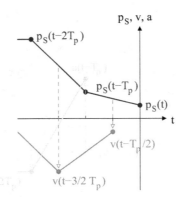

Figure 6.8: Estimating acceleration from periodical position fixes $p_S(t)$ every T_p obtained from the GPS.

as secant slopes of subsequent position fixes. With T_p denoting the nominal period between two subsequent position fixes from the GPS and $p_S(t)$ representing a position measurement with timestamp t, this calculation of velocity and acceleration estimates can be formalized as follows:

$$v(t - T_p/2) \quad = \quad \frac{p_S(t) - p_S(t - T_p)}{T_p} \tag{6.17}$$

$$a(t - T_p) \quad = \quad \frac{v(t - 3/2 \cdot T_p) - v(t - T_p/2)}{T_p} \tag{6.18}$$

Thus, the acceleration estimate is delayed by T_p relative to t. Figure 6.8 depicts this situation based on graphs of position, velocity, and acceleration.

Therefore, inserting $a(t - T_p)$ into (6.2) yields a good estimation of the "true" gravity vector. However, $a(t - T_p)$ is measured by the GPS in world coordinates (BCS), so it must be transformed to VCS. Formally:

$$
\begin{aligned}
{}^{VCS}g^*(t - T_p) \quad &= \quad {}^{VCS}a_S(t - T_p) + {}^{VCS}a(t - T_p) \\
&= \quad {}^{VCS}a_S(t - T_p) + {}^{BCS}VCS^{\mathrm{T}}(t - T_p) \cdot {}^{BCS}a(t - T_p) \tag{6.19}
\end{aligned}
$$

The MARVIN algorithm uses the past attitude filter output at $t - T_p$ for ${}^{BCS}VCS(t - T_p)$ in (6.19). This is a simplification, because it is actually the error in ${}^{BCS}VCS(t - T_p)$ that is sought for. However, since the sum in (6.19) is dominated by gravity, the effect of the former orientation error on the resulting g^* estimation is always lower than the former orientation error itself, which actually assures convergence. Please note that a more formally correct determination of g^*, say through solving a system of equations for ${}^{BCS}VCS(t - T_p)$, would still suffer from the error in the secant approximation of $a(t - T_p)$, which may be in a similar order of magnitude.

The past magnetic field sensor reading $B_S(t - T_p)$ yields a second reference vector:

$$
{}^{VCS}N^*(t - T_p) \quad = \quad {}^{VCS}B_S(t - T_p) \tag{6.20}
$$

Now, these two vectors g^* and N^* do fulfill the requirements on the application of (6.13–6.16), so a "corrected" past attitude frame ${}^{BCS}VCS^*(t - T_p)$ is calculated via these equations. By expressing both this corrected reference attitude and the past filter output ${}^{BCS}VCS(t - T_p)$ as a vector of yaw, pitch, roll Euler angles according to (5.108), a "calibration error" β is obtained as:

$$\beta(t) = \begin{pmatrix} \phi^* \\ \theta^* \\ \psi^* \end{pmatrix}(t - T_p) - \begin{pmatrix} \phi \\ \theta \\ \psi \end{pmatrix}(t - T_p) \tag{6.21}$$

Here, $\beta(t)$ is associated with the time of its computation, but it still refers to the past time $t - T_p$.

Before using $\beta(t)$ for correction purposes, it is first low-passed through a discrete approximation of a first-order low-pass filter:

$$\dot{\overline{\beta}}(t) \approx \lambda_1 \left(\beta(t) - \overline{\beta}(t) \right) \tag{6.22}$$

$$\overline{\beta}(t + T_f) = \overline{\beta}(t) + \lambda_1 T_f \left(\beta(t) - \overline{\beta}(t) \right) \tag{6.23}$$

Here, λ_1 approximates the angular cut-off frequency ω_e of this low-pass, and T_f denotes the period of the filter loop. The application of this low-pass allows to reduce the effect of noise from the accelerometer and from ${}^{BCS}a$ estimation on the filter output while maintaining a comfortably high compensation rate in case of significant persistent drift of the integrated rotation sensor readings.

Both $\overline{\beta}(t)$ and the current rotation rate sensor output $\omega_S(t)$ demand for a rotation of the state vectors ${}^{VCS}g$ and ${}^{VCS}N$. Computation time can be saved and rounding errors reduced by combining both components into a single vector α of rotation angles:

$$\alpha(t) = T_f \cdot \omega_S(t) + \lambda_2 T_f \cdot \overline{\beta}(t) \tag{6.24}$$

Indeed, this is again a simplification, because $\overline{\beta}$ does *not* refer to the current VCS axes as of t, but instead consists of averaged Euler angle errors of some time in the past. At this point, the filter algorithm presumes some linearity of rotational transforms that does not truly exist, but is a good-enough approximation in case of a "small" calibration error $\overline{\beta}$, which in turn is maintained through the filter algorithm. This simplification in some way corresponds to the local linearization that is performed by an extended Kalman filter for coping with nonlinear dynamics.

For numerical reasons set forth in the next section, the final state update is performed according to the approximate law

$$ {}^{VCS}g(t + T_f) = Rot_{P(0)}(\alpha^{P(0)}) \cdot Rot_{P(1)}(\alpha^{P(1)}) \cdot Rot_{P(2)}(\alpha^{P(2)}) \cdot {}^{VCS}g(t) \tag{6.25}$$

$$ {}^{VCS}N(t + T_f) = Rot_{P(0)}(\alpha^{P(0)}) \cdot Rot_{P(1)}(\alpha^{P(1)}) \cdot Rot_{P(2)}(\alpha^{P(2)}) \cdot {}^{VCS}N(t) \tag{6.26}$$

with $P : \{0; 1; 2\} \rightarrow \{x; y; z\}$ some *random* permutation of the axes, drawn anew in every computation cycle. This imposes an arbitrary random order on the rotations that should actually occur concurrently.

Parameter	Symbol	Value
filter period	T_f	1/40 s
GPS position fix period	T_p	1/10 s
filter gain for $\bar{\beta}$	λ_1	2.500 s^{-1}
filter gain for α	λ_2	0.625 s^{-1}

Table 6.1: List of parameters of the attitude filter algorithm as used in the MARVIN system.

λ_2 in (6.24) is actually the gain of another low-pass filter that adjusts the compensation rate of the calibration error. Assuming the first low-pass from (6.23) already converged, the upper bound on the filter gain from (5.47) holds here as well, with T_p serving as dead time and λ_2 as filter gain, indicating that divergence of the filter will certainly occur unless:

$$\lambda_2 \; < \; \frac{\pi}{2T_p} \tag{6.27}$$

The combination of (6.23,6.24) actually implements a second-order low-pass on the calibration error $\beta(t)$.

In order to compensate for accumulated errors resulting from the periodic application of the update rule (6.25,6.26), the following renormalization is applied occasionally. It adjusts the vectors to unit length and ensures that they are perpendicular:

$$^{VCS}g(t)_{\text{norm}} \; = \; \frac{^{VCS}g(t)}{\|^{VCS}g(t)\|} \tag{6.28}$$

$$^{VCS}N(t)_{\text{norm}} \; = \; \frac{^{VCS}N(t) - (^{VCS}g(t) \cdot {}^{VCS}N(t)){}^{VCS}g(t)}{\|^{VCS}N(t) - (^{VCS}g(t) \cdot {}^{VCS}N(t)){}^{VCS}g(t)\|} \tag{6.29}$$

The implementation in the MARVIN system uses the parameter settings listed in table 6.1. Please note the big safety margin in the setting of λ_2, with the limit imposed by (6.27) at 15.7 s^{-1}.

6.3.2 Analysis

In this section, the fusion algorithm described in 6.3.1 is analyzed for the effects of the deliberately chosen simplifications and for its accuracy and computational cost compared to an extended Kalman filter as the common "state-of-the-art" method.

Simplifications

First, all computations in the MARVIN filter implementation are performed in 32 bit fixed-point arithmetics, because the on-board SAB80C167 microcontroller [12] does not posses an FPU and floating point calculation in software results in high computational and memory cost (the latter mainly for the space occupied by library routines). Due to the necessary multiplications, this restricts the resolution of all intermediate results to 16 bit (signed), i.e. the

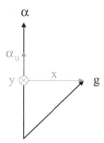

Figure 6.9: Approach for implementing "correct" rotation about angle vector α.

magnitudes' resolution to 15 bit, at most. This has important consequences on the desirability of certain *qualitative simplifications*, as will turn out later.

Qualitative simplifications, as involved in the filter equations from section 6.3.1 aside from numerical inaccuracy, are constituted by the following four issues:

1. Acceleration estimation according to (6.18) and figure 6.8 gives only an average over an interval, not a point-of-time measurement.

2. The transform of motion acceleration into the VCS according to (6.19) uses a (slightly) incorrect VCS orientation.

3. The combination of the measured rotation and the calibration error according to (6.24) uses (slightly) incorrect axes for the calibration term.

4. State vector rotation according to (6.25,6.26) is mathematically different from the actual "concurrent" rotation about $\alpha(t)$ by its magnitude.

Simplification 1 is somewhat inevitable and about the best thing one can do considering the nature of the position data available. Simplification 2 has already been discussed above; its avoidance would *not* lead to a significantly reduced error in the presence of the previous one. Simplification 3 takes the role of local linearization on the one hand, as already explained above, and will appear particularly reasonable on the other hand after the analysis of the last simplification, since it minimizes the number of rotations to be actually performed.

Justifying simplification 4 requires more reflection. For performing a "correct" rotation of $g \in \mathbb{R}^3$ about an angle vector $\alpha \in \mathbb{R}^3$, a highly optimized implementation with regard to the number of arithmetic operations performed is illustrated by figure 6.9: The rotation occurs in a plane spanned by vectors x and y and containing the point $g - x$. Formally, the image vector g' is then obtained through:

$$\alpha_u \;=\; \alpha \,/\, \|\alpha\| \tag{6.30}$$

$$y \;=\; \alpha_u \times g \tag{6.31}$$

$$x \;=\; y \times \alpha_u \tag{6.32}$$

$$g' \;=\; g - x + x \cos\|\alpha\| + y \sin\|\alpha\| \tag{6.33}$$

Rotation Scheme	multiply	divide	square root	sin/cos
"correct" α rotation (6.30–6.33)	3	1	1	1
subsequent x, y, z rotations (6.34–6.36)	2	0	0	1

Table 6.2: *Maximum* number of *subsequent* truncating operations for "correct" and simplified implementations of state vector rotation.

Rotation Scheme	multiply	divide	square root	sin/cos
"correct" α rotation (6.30–6.33)	18	1	1	2
subsequent x, y, z rotations (6.34–6.36)	12	0	0	6

Table 6.3: *Total* number of truncating operations for "correct" and simplified implementations of state vector rotation.

On the other hand, "simplified" subsequent rotation by the three components of α, as defined in (6.25,6.26), would result in the following sequence of operations, shown here for the permutation x, y, z:

$$g_1 = \text{Rot}_z(\alpha^z) \cdot g = \begin{pmatrix} g^x \cos \alpha^z - g^y \sin \alpha^z \\ g^x \sin \alpha^z + g^y \cos \alpha^z \\ g^z \end{pmatrix} \quad (6.34)$$

$$g_2 = \text{Rot}_y(\alpha^y) \cdot g_1 = \begin{pmatrix} g_1^x \cos \alpha^y + g_1^z \sin \alpha^y \\ g_1^y \\ -g_1^x \sin \alpha^y + g_1^z \cos \alpha^y \end{pmatrix} \quad (6.35)$$

$$g' = \text{Rot}_x(\alpha^x) \cdot g_2 = \begin{pmatrix} g_2^x \\ g_2^y \cos \alpha^x - g_2^z \sin \alpha^x \\ g_2^y \sin \alpha^x + g_2^z \cos \alpha^x \end{pmatrix} \quad (6.36)$$

The accuracy of the result is adversely affected by fixed-point arithmetics whenever an intermediate or final result needs to be truncated. All values during computation are represented using 16 bit so that multiplication of two values is always possible with a 32 bit result. Hence, rounding to 16 bit must occur after every multiplication. The maximum number of these truncations occurring *subsequently* on any component of the inputs g and α to yield some component of the result vector g' thus provide a reasonable measure of the accuracy to be expected.

Table 6.2 lists this number of subsequent truncating operations for the two implementations shown above. Truncating operations, i.e. operations that cannot be performed 100% accurately in fixed-point arithmetics, are multiplication followed by 16-bit truncation, division, square root, and trigonometric functions. The squares involved in the computation of $\|\alpha\|$ are not considered because they do not require rounding. While the number of *subsequent* truncations is most relevant for effective accuracy, table 6.3 lists the *total* number of truncating operations required by each of the algorithms, for completeness.

The tables show that multiplication is the primary source of rounding errors in both implementations. The "correct" rotation, with one division and three multiplications in sequence, is indicated to be prone to more rounding error than the "simplified" one, which only requires two subsequent multiplications since each rotation changes only two of three vector components. The simplified implementation, however, exceeds the alternative one with regard to six sin/cos functions to be computed in total compared to only two.

To verify this code-level comparison on the basis of practical performance, simulations of the two algorithms have been carried out on a PC to evaluate the accumulating error, with proper α-rotation computed in 64 bit floating-point serving as reference. Figure 6.10 shows the accumulating angle error (angular deviation between fixed-point and reference rotated vectors) and the accumulating length error (length deviation of fixed-point rotated vector relative to the reference rotated vector's length) for the two algorithms. In this experiment, 3000 subsequent rotations about 7 different angles between $0.03°$ and $1.6°$ per axis (average of $0.50°$) are performed. This range of rotation angles is in the order of magnitude found in the MARVIN case. The angles used in this experiment are only an example, of course, but the result is not qualitatively different for other sets of angles. The figure shows that the "mathematically incorrect" implementation actually yields better results. While the length errors are similar and slightly lower for the "correct" algorithm (0.9% vs. 1.6%), the angular error of the "simplified" implementation is lower by a factor of almost 5 ($0.6°$ vs. $2.9°$).

As the filter examined here is an attitude filter and the filter output according to (6.13–6.16) does not respond to length errors at all, the "simplified" algorithm is to be preferred, as already indicated by (6.25,6.26). The error caused by "incorrect" sequential rotation is clearly dominated by the rounding errors occurring in the more sophisticated computation. The different observations for angular and length errors may be explained as follows:

- The "correct" algorithms suffers most from the computation of $\|\alpha\|$. For small angles, it loses resolution from the fact that only a single sin/cos term, respectively, is evaluated, its argument subject to rounding involved in the computation of the square root. The "simplified" algorithm computes the sin/cos factors directly from the vector components instead.

- The sin/cos multipliers, on the other hand, are primarily responsible for length errors in the image vector, so that the reduced number of sin/cos multipliers applied in the "correct" algorithm seems to improve its preservation of the vector's length.

This observation also justifies simplification 3 listed above, by suggesting that the use of slightly incorrect rotation axes might be less harmful than the saving of one half of the required rotation operations beneficial. All in all, it is a good demonstration of the fact that formal exactness should not necessarily be the primary goal during small UAV system design. Instead, various effects need to be carefully balanced in a quantitative way to finally obtain the best possible trade-off of performance and effort.

Accuracy

To obtain an absolute benchmark of the accuracy and suitability of MARVIN's attitude filter, an extended Kalman filter (EKF) has been implemented as a reference on a PC. The work

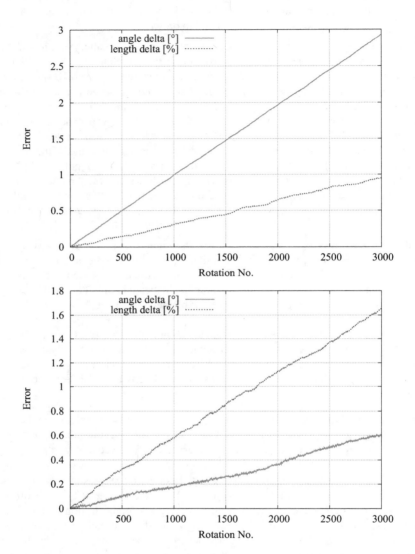

Figure 6.10: "Correct α rotation" (top) and "subsequent x, y, z rotation" (bottom) errors using 32-bit fixed-point arithmetics.

reported in this section has been primarily conducted by Carsten Deeg and jointly published in [105]. Further details of the underlying EKF model are taken from [38].

The filter implementation follows the Kalman-Bucy filter [83] algorithm for continous systems together with numerical Runge-Kutta integration over time. The update of the state vector is performed using the standard EKF approach with Jacobians describing local linearization.

The filter uses the following state description S,

$$S = \begin{pmatrix} ^{BCS}p \\ ^{BCS}v \\ ^{BCS}a \\ \psi, \theta, \phi \\ ^{VCS}\omega \end{pmatrix} \in \mathbb{R}^{15}, \tag{6.37}$$

which is applicable to any solid object moving with 6 degrees of freedom. It consists of the following components:

- $^{BCS}p \in \mathbb{R}^3$ is the position of the vehicle,

- $^{BCS}v \in \mathbb{R}^3$ is the velocity of the vehicle,

- $^{BCS}a \in \mathbb{R}^3$ is the acceleration of the vehicle,

- $\psi, \theta, \phi \in \mathbb{R}$ are the usual Euler angles describing the vehicle's attitude, as defined through (5.108),

- $^{VCS}\omega \in \mathbb{R}^3$ is the current rotation of the vehicle.

As the filter does not use any information about helicopter dynamics, the control input is completely neglected. Therefore, the prediction stage relies on trivial rotation and translation laws only, presuming zero change rates of both ^{BCS}a and $^{VCS}\omega$.

The measurement stage uses a measurement vector Y,

$$Y = \begin{pmatrix} ^{BCS}p_S \\ ^{BCS}v_S \\ ^{VCS}a_S \\ ^{VCS}B_S \\ ^{VCS}\omega_S \end{pmatrix} \in \mathbb{R}^{15}, \tag{6.38}$$

consisting of

- $^{BCS}p_S, ^{BCS}v_S \in \mathbb{R}^3$, the position and velocity measurement obtained from the GPS,

- $^{VCS}a_S \in \mathbb{R}^3$, the accelerometer measurement,

- $^{VCS}B_S \in \mathbb{R}^3$, the magnetometer measurement, and

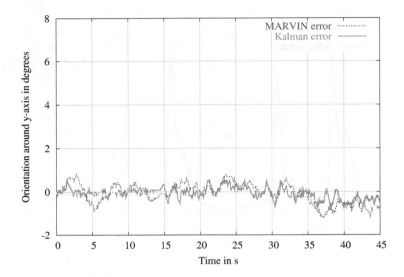

Figure 6.11: Error of VCS-*y* orientation from MARVIN's attitude filter and EKF together with true orientation.

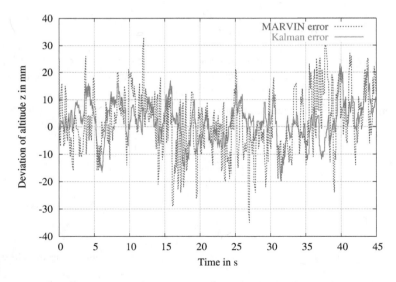

Figure 6.12: Error of BCS-*z* position from MARVIN's extrapolation and EKF.

- $^{VCS}\omega_S \in \mathbb{R}^3$, the measurement obtained from the rotation rate sensors.

There are no further simplifications of the Kalman algorithm involved. In the implementation of the algorithm, a 15×15 Jacobian matrix has to be inverted, which is done numerically.

Figure 6.11 shows the resulting errors in the VCS-y (actually, θ) orientation estimation inferred by MARVIN's filter algorithm described in 6.3.1 and the EKF. These plots have been created using a simulated flight and emulating the sensors with simulated noise. Together with the two filter errors, the actual orientation is shown. The peaks in the orientation indicate flight maneuvers involving positive acceleration in the direction of VCS-x.

Figure 6.12 depicts the same comparison for the vertical position ($^{BCS}p^z$) of the helicopter. MARVIN's position filter uses simple linear extrapolation of GPS position data based on the last available position fix and the last available velocity estimation according to (6.17).

Comparing the errors from the two filter algorithms, it turns out that they are of the same order of magnitude. At best, one can state that the worst-case deviations of the EKF are better by a factor of 0.5. However, the EKF outputs are significantly less noisy, but even the higher noise level of the MARVIN filter is far from being harmful.

On the other hand, the computational effort of computing the EKF, with possible compiler optimizations enabled, has been measured to be roughly 200 times higher (890 μs vs. 4.5 μs on a P4 2.8 GHz and gcc 3.3.1, see [38]). Since the microcontroller aboard MARVIN does not have an FPU, the use of the EKF is clearly out of the question, given that the simple MARVIN filter clearly performs well enough.

Furthermore, the sensors in the simulation have been emulated with exactly the same white noise as modeled in the EKF. This means that the performance of the EKF should be expected to degrade in an actual application.

6.3.3 Failsafe Procedures

For maximum safety in case of component failure, the implementation of dedicated failsafe procedures for attitude filtering is highly desirable. While Kalman-filter-based solutions tend to require the implementation of separate filter modes for every kind of partial sensor failure, the design of failsafe procedures for specialized algorithms like the MARVIN filter presented above is more straightforward. This section proposes failsafe procedures for the MARVIN attitude filter.

GPS failure requires setting $^{VCS}a = 0$ in (6.19). But effectively, this completely disables gyro drift compensation, as set forth in section 6.1. Hence, only failures of short duration (few seconds) can be bridged without manual intervention.

The failure of accelerometers and/or magnetic field sensors should be addressed by disabling gyro drift compensation completely, i.e. by setting $\overline{\beta}(t) = 0$ in (6.24). This, also, will require manual intervention unless the failure ceases after a few seconds, if rotation rate sensors of low accuracy are employed.

The failure of rotation rate sensors is the worst case possible. No reasonable failsafe procedure exists and *immediate* manual intervention will be required.

6.4   Image Sensors

The term *image sensors* shall embrace all kinds of visual light, infrared, and ultraviolet cameras, and both video and still-picture camera types. Anyway, with digital image processing, the discrimination between still-picture and video sources has become largely obsolete. For still-picture cameras can of course output a sequence of frames as well, and video cameras do differ widely in their possible frame rate. Therefore, the classical binary discrimination needs to be replaced by a listing of descriptive technical features, the relevant ones in this respect being supported *resolutions*, *frame rates*, *coding schemes*, and *interfaces*.

Image sensors are probably the most frequently desired small UAV sensors targeted at mission data acquisition. This fact and the great variety of available camera types, as already indicated by the consideration in the previous paragraph, permit to deal with the class of image sensors in a general way, without having to assume any special type of mission. This is the purpose of the current section.

The remainder of this section, in seeming contradiction to this introduction, is structured according to the obvious perspective of "shopping" into the discussion of analog video cameras, digital video cameras, and finally digital photo cameras.

6.4.1   Analog Video Cameras

The only cameras with non-digital output to be considered in this book are analog video cameras with an analog electric video signal output[3].

Analog video cameras in combination with framegrabber hardware have been the "favorite" data source in digital image processing for quite some time. Today, with digital video cameras available (see next section), this combination has lost most of its attraction. For an unnecessarily long analog signal path is susceptible to signal distortions, and framegrabber hardware tends to be overly expensive and unnecessarily restrictive in the choice of a suitable computing platform.

However, for use aboard *truly* small UAVs, miniaturized analog video cameras in combination with an analog radio transmission link are available at very low cost.

One such camera-transmitter set, available from Conrad, Germany [120], shows the following set of relevant features:

- CMOS color camera, 628x562 pixels (PAL),
- transmission frequency 2.4 GHz,
- camera power consumption 7.5 V, 300 mA,
- transmission range up to 100 m outdoors, 30 m indoors,

---

[3]Carrying analog film cameras aboard a UAV is an option as well but only for highly specialized mission types, which is why it will not be considered beyond this footnote.

- receiver output PAL composite video,

- camera size (without stand) 26 mm × 26 mm × 27 mm,

- camera mass (without stand) 40 g,

- price (camera + receiver) 150 EUR.

Of course, analog video transmission leads to low image quality and usually lots of visible interference. The limited transmission range poses another restriction on the application of cameras like this. Nevertheless, it may be the only option available today for acquiring image data from an ultra-small UAV without on-board image processing capabilities.

6.4.2   Digital Video Cameras

Digital "video" cameras are mainly available today with one of the following interfaces:

- *FireWire*, or IEEE 1394 [127], is an interface commonly used for digital video stream transmission. The 1995 version provides up to 400 Mbit/s data rate. Since FireWire has been designed with video stream transmission as one of its primary applications, it is the most "natural" replacement of analog video wiring today. Many cameras with FireWire interface support a standardized camera-to-host interconnect [17] according to IIDC, a working group within the *1394 Trade Association*. These cameras may be used even if no specialized camera driver is available for the on-board computing platform, but only a general "IIDC" driver.

- USB (see 3.1.2) is also suitable for digital camera interfacing. It is very commonly used in webcams, since every PC is equipped with a USB interface today. If the on-board computer can act as USB host, using a USB webcam is a reasonable low-cost, low-weight solution. Of course, image quality will be rather poor, but designers can benefit from the huge sales figures of these devices with respect to size and cost. Specifications are subject to almost daily changes, but cameras with 15 frames per second (fps) at 640 × 480 pixels weighing about 20 g can be purchased for around 10 EUR. However, the necessary implementation of a suitable device driver, unless standard PC architecture is used on board, may constitute an obstacle.

- Recently, many *network cameras* have become available. These can be directly connected to an Ethernet cable. Usually, they even embed a web server for remote access and configuration, but the use as an on-board camera should not be difficult typically. VGA resolution (640 × 480) and 30 fps are available below 200 g and at very moderate cost. However, due to their primary purpose as stand-alone webcams, they might output images in JPEG-compressed format only.

6.4.3   Digital Photo Cameras

Designers of robotic systems have been exhibiting a tendency to shop for "video" cameras ever since. However, with respect to image quality, digital *photo* cameras are far ahead of their "video" competition, even more so when considering cost and size as well.

Probably all quality cameras can be both operated and the images read out via their host interface. Today, this interface is usually USB, occasionally FireWire as well. But unfortunately, these cameras are not at all targeted at embedded applications. Therefore, it is difficult to obtain software or documentation permitting this mode of operation.

Particularly annoying is that camera manufacturers tend to treat the communication protocols implemented in their products as "classified information". They consistently provide drivers for the Windows and Mac OS operating system families only. gPhoto, for example, the open-source project focussed on camera interfacing, has been forced to reverse-engineer almost every camera protocol on the market, see [3]. Their website reports that several manufacturers have openly declined to support this project, although providing support at this place might indeed have affected a certain number of customers in their preference of a camera brand. This is particularly incomprehensible because of the fact that enterprises like gPhoto do not have too much difficulty in reverse-engineering these protocols, since the communication between a camera and a host computer can easily be eavesdropped and analyzed. If some open-source developers can spend the effort to do so, any competitor would only laugh about this kind of secrecy.

It is, however, possible in some cases to obtain a *software development kit* (SDK) or library that makes the camera accessible to software written by the customer. Manufacturer policies do vary in this respect. The MARVIN team has once signed a dedicated NDA for requesting such a software development kit and sent this document to Nikon's Japan headquarters, but without getting any response. Fortunately, Canon do provide their Digital Camera SDKs after an on-line application within five business days – to business entities at least [2]. The Canon SDK is used with good results in the MARVIN system, which is equipped with a "vintage" Canon PowerShot S45 [115, 68]. However, this SDK is constantly updated to support all cameras manufactured by Canon, even including their digital SLR series. However, the SDK is, as "usual", only available for the Windows and Macintosh platforms.

The predecessor of the current MARVIN UAV, "MARVIN Mark I", which performed its first autonomous flights in 1999, was equipped with an early digital photo camera manufactured by Samsung, a VPC-X350EX [100, 102]. This camera has an RS 232 interface instead of USB, by which it was connected to the on-board microcontroller. Due to the data rate limit of the camera's interface of only 230 kbit/s, the minimum frame interval time was about 6 s for images of only 640 × 480 pixels. Nevertheless, the choice of a digital photo camera for image acquisition turned out to be one critical aspect in winning the IARC final in 2000 [101]: the resulting image quality, as indicated by the sample images presented in figure 6.13, was by far superior to all competing UAVs[4], which used analog image transmission without exception. For the Samsung camera, a member of the MARVIN team, Volker Remuß, took care of reverse-engineering the serial protocol through eavesdropping.

---

[4]To all teams *that had already addressed* image processing, that is.

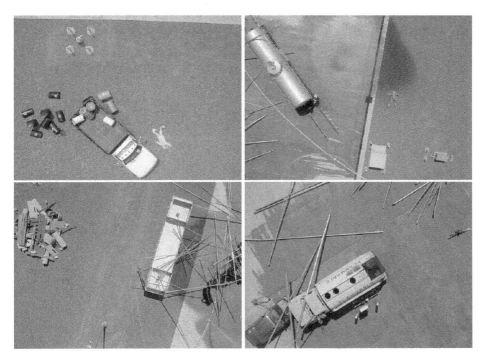

Figure 6.13: On-board photos from "MARVIN Mark I" taken during the 2000 IARC final.

Figure 6.14: On-board photo from MARVIN with magnified crop.

The Canon PowerShot S45 aboard the most recent MARVIN "Mark II" UAV [115, 68] is connected via USB to an embedded PC that has the Canon SDK installed. Even with the faster interface, images in the camera's maximum resolution of about 4 megapixels require about 8 s for pipelined read-out and transmission. But patience is rewarded with exceptional quality. Figure 6.14 shows an in-flight picture taken by the MARVIN UAV with an overlaid crop magnification that shows the readability of the cars' license plates. Alternatively, storing images on the camera's memory card permits up to 1 fps of sustained shooting in full resolution.

A very useful feature of the PowerShot S45 in connection with the Canon SDK is the so called *viewfinder mode*: In this mode of operation, the camera outputs only 320 × 240 pixels, but at about 8 fps sustained frame rate. This corresponds, both in resolution and update rate, to the data stream optionally displayed on the LCD screen on the back of the camera. Having this small video stream is probably sufficient for any sort of dynamic human interaction from the ground station and may finally dispel any concern with regard to the alleged necessity of a true "video" camera.

6.5 Subsumption

This chapter about on-board sensors has pointed out information of relatively general relevance. The first major aspect consists in the interdependence between different sensor types with respect to attitude measurement. In general, attitude measurement without a working position sensor is not possible with low-cost sensors as probably used in small UAV systems. Furthermore, accelerometers and magnetometers – or a suitable model of the vehicle dynamics – are required for the on-line correction of integrated rotation rates.

A second major issue is the insight that sensor fusion filters for attitude measurement are not per se difficult to implement. However, when performance and cost trade-offs are critical, it pays off to examine the *quantitative* effects of all sorts of design decisions very closely. Off-the-shelf methods like Kalman filters may be used, but need not be preferable in every case.

Finally, when selecting an on-board camera, more options than obvious at first glance may be worth considering. If image quality constitutes a vital requirement, a digital photo camera might not only be superior to "video" equipment in this respect, but also cheaper.

Chapter 7

Development and Safety Tools

This chapter discusses issues related to the design process and operation safety of a small UAV system. These are *not necessarily* part of the requirement specification to be met through the design. Instead, these tools support various stages of the development process and shall ensure maximum possible safety in the application phase. Such tools may also have a critical impact on the duration and success of the development process, as will be motivated in the following sections.

The tools to be discussed in one section each are software-in-the-loop simulation (7.1), data logging and analysis (7.2), and failsafe procedures (7.3). This chapter closes, too, with a short subsumption.

7.1 Software-in-the-Loop Simulation

In the development of embedded systems, two procedures known as *software-in-the-loop* simulation and *hardware-in-the-loop* simulation (e.g. [96]) are common for testing the software and/or hardware under development. While the environment is simulated in software in both cases, the communication between this simulated environment and the embedded software under test works differently in each case:

- In the case of *hardware-in-the-loop* simulation, the embedded system (e.g. the UAV) is operated "as usual", but actuator outputs and sensor inputs are recorded or provided, respectively, through special emulation hardware. This approach is depicted by figure 7.1.

  Of course, certain parts of the hardware may have to be left out, e.g. the engine would not necessarily run in the case of a helicopter UAV. Acceleration and rotation measurements could be approximately provided by rotating the vehicle in a cardanic frame, for example.

- In the case of *software-in-the-loop* simulation, only the software normally running in the embedded system is tested. All hardware involved is replaced by a special emulation software layer that reads out or injects data at the lowest possible software interface level. Figure 7.2 visualizes this approach.

207

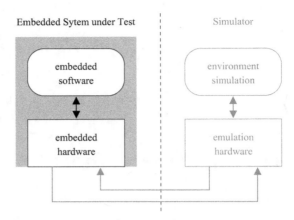

Figure 7.1: Schematic illustration of system test via *hardware-in-the-loop* simulation.

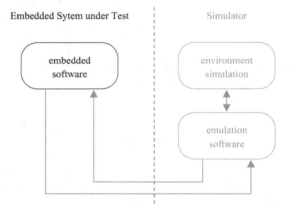

Figure 7.2: Schematic illustration of system test via *software-in-the-loop* simulation.

In the case of sensors connected through analog-to-digital converters, one would re-place the reading of the converter's data register by reading a variable that reflects the simulated value of the corresponding signal. In the case of an "intelligent" sensor that communicates via an RS 232 or USB interface, the sensor's data log would be created by the emulation software layer instead and injected into some data buffer that is read by the embedded software under test. The emulation layer gains access to the embedded software's outputs in an analog way.

Due to the tight coupling between embedded and emulation software and the difference in operation compared to normal operation, pure software-in-the-loop tests are most often performed on a platform different from the embedded processor.

Obviously, the hardware-in-the-loop approach cannot be followed to full extent usually. GPS, for example, would require simulated satellites (*pseudolites*) for transparent hardware-in-the-loop testing. Magnetic field emulation is also very complicated and expensive [96], exact emulation of acceleration is completely impossible. Hence, hardware-in-the-loop simulation environments will usually have to revert to software-in-the-loop simulation for certain parts of the system under test. The GPS receiver, for example, could be transparently replaced by an external computer outputting the receiver logs via an RS 232 connection instead of the actual GPS unit. This results in an effective mixture of the two alternative approaches.

Apart from this problem with the hardware-in-the-loop approach, there are a couple of other reasons that clearly favor software-in-the-loop simulation as a recommended general tool in small UAV development:

Effort. Clearly, the creation of a testing environment for "true" hardware-in-the-loop simulation is complex, expensive, and time consuming. In the prototype-phase of system design, it may be difficult to afford.

Utility. When examining complex embedded systems, it is evident that complexity arises from the combination of system components, which in turn is only effectively reflected on the software level. It is comparatively easy to design a single sensor or to connect it to the embedded system because a task like this can largely be performed and tested in isolation. Usually, these low-level component interfacing issues are addressed early in the design process and quite rarely require redesign steps. Thus, the great majority of relevant issues will occur in the software layer, where they are accessible through software-in-the-loop tests as well.

Unambiguity. When faulty behavior is detected during testing, one must always determine if the fault occurred in the simulation environment or in the system under test. This, of course, holds for both testing approaches, but testing environments that are purely software are less prone to wear, damage, interference, and other kinds of spontaneous failure. Thus, the results obtained in software-in-the-loop testing can be considered less ambiguous.

Flexibility. Proper hardware-in-loop testing is only possible in the laboratory that contains the testing environment. Software-in-the-loop tests, in contrast, may be performed on any computer, possibly even on a laptop while travelling, so that the number of concurrently active tests and testers and their whereabouts are virtually unlimited. In particular, software-in-the-loop testing may also be possible "in the field", while actually changing and testing properties of the *real* system in rapid succession.

Speed. Fully transparent hardware-in-loop testing must take place in real time, that is, take as much time as some test-case would with the real system. Software-in-the-loop testing, on the other hand, may be based on simulated clocks and run with the maximum speed of the computing platform in use. Hence, test-cases of minutes or hours of simulated real time may be finished in a couple of seconds.

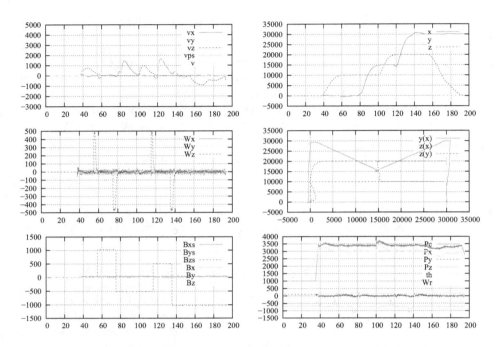

Figure 7.3: Sample result from software-in-the-loop testing with MARVIN.

Therefore, the hardware-in-the-loop approach seems only desirable in the presence of very special requirements – like in the final phase of product development, or if situations need to be examined that are not realizable during real field tests. Otherwise, software-in-the-loop testing will be more beneficial, and should actually be implemented in every case of UAV development.

Figure 7.3 shows a sample plot from software-in-the-loop testing with the MARVIN system. Details regarding the methods used can be found in [38]. The six diagrams document a simulated flight of 195 s, plotting a number of different variables. While the details of this test-case are not of interest here, a very short description of each diagram is provided as an overview. Position is given in mm, time in s, and angle in $1/10°$ (approximately), with the derivatives' units accordingly. From top left to bottom right, row by row, the information plotted is:

1. velocity in BCS-x, y, z, desired velocity, and magnitude of velocity over time;

2. position in BCS-x, y, z over time;

3. rotation rate in HCS-x, y, z over time;

4. position in BCS-x, y, z as projections y over x, z over x, and z over y;

5. desired and actual Euler angles ψ, θ, ϕ over time;

6. servos control signals and main rotor RPM over time.

This test-case includes autonomous take-off and landing, plus mode-switching in the controller and curve flight. On a rather outdated P4 1.6 GHz, the runtime of the simulation is only 4.6 s[1].

7.2 Data Logging and Analysis

Both during development and application, data logging features are desirable that record the developing of many state variables for later analysis. The general utility of variable logging is beyond doubt for any embedded system. In the case of UAV systems, data logging facilities can be classified according to two independent discriminations:

- The *location* of the data log. This can be either some data storage device aboard the vehicle, or be performed at the ground station.

- The *purposefulness* of the data log. On the one hand, data logging may be performed for a specific set of variables in order to answer a particular question that has arisen beforehand. One example would be a dedicated flight for the identification of model parameters, as discussed in section 5.5.3. On the other hand, data logging may be performed routinely according to the idea of a *flight data recorder* (FDR), the "black box" obligatory in air transportation[2].

The optimal location of the data log is, of course, the ground station. For first, data analysis will be performed by human operators in the ground and should be possible without physical access to the UAV, and second, data stored aboard may be lost in a crash of the UAV, which in turn is exactly the situation where these data are most significant. However, having to transmit telemetry data to the ground station may impose an undesirable limit to the possible sampling rate. Therefore, combined solutions involving temporary storage aboard during the acquisition of data logs at high sampling rates and the subsequent wireless transmission of these logs to the ground station may constitute the best compromise in the face of such requirements.

Anyway, such combined solutions of airborne *and* ground storage are only applicable to data logging explicitly triggered for a particular purpose, which is why FDR-style data logging must be done with a sampling rate low enough to permit sustained transmission to the ground station. FDR-style logging should not be neglected in any case, because experience tells that at some time there will *always* be questions that arise *after* the flight, maybe much later.

In the MARVIN system, FDR-style logging at the ground station is performed with peak sampling rates of about 5 Hz. This holds for all raw sensor outputs and both intermediate

---

[1]This does *not* include full-featured aerodynamic simulation of the main rotor according to blade-element momentum theory, see [38] for details. Instead, the induced speed of the airflow through the rotor disc is assumed to be homogeneous. The results of this approach are still good enough to probably reveal any software fault that would be hazardous to the real helicopter in the same flight situation.

[2]The analog of the second type of "black box", the *cockpit voice recorder* (CVR), would – if at all – have to be provided at the ground station in the case of UAV operation.

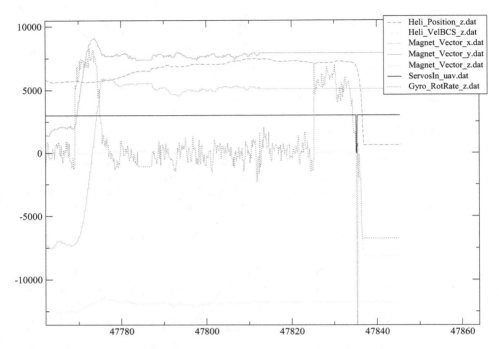

Figure 7.4: "FDR plot" unveiling the causes of (the only ever) crash during autonomous flight.

results and outputs of the position and attitude filters. The infrastructure for the transmission of these variables at a guaranteed minimum bandwidth is optimally provided by the BBCS communication network, see section 4.4. Of course, 5 Hz is too slow for isolating certain kinds of noise and interference. But nevertheless, these data logs still permit to retrace the vast majority of aspects of closed-loop behavior.

Figure 7.4 shall solve as an example. It depicts the last 74 s of autonomous flight of a third-party helicopter UAV running the MARVIN software for its autonomy, this flight ending with the only crash ever experienced during autonomous flight with the MARVIN systems. During hovering, the UAV started to drift away and lose altitude all of a sudden. The human safety pilot tried to switch back to manual operation to regain control, but the helicopter did not respond to the pilot's remote control commands.

The figure shows seven variables plotted over time in s, approximately $t \in [47762\,\text{s}; 47836\,\text{s}]$. *Heli_Position_z* and *Heli_VelBCS_z* are BCS-z position and velocity, respectively, in mm and mm/s. The abrupt loss of altitude starting at $t \approx 47835$ s is clearly visible, with the magnitude of vertical speed at the time of impact, $t \approx 47837$ s, reaching 8.1 m/s. Only with the help of the data log depicted here, the primary cause of the crash could be attributed to the failure of the on-board compass, visualized through its raw field vector outputs *Magnet_Vector_x,y,z*. The plots show that the field vector stops changing at $t \approx 47813$ s, more than 20 s before the observable event. This did not cause an immediate problem, because the helicopter was hovering with constant yaw angle, and the short-term output of the attitude filter relies on the rotation rate sensors, anyway. A prior change in the yaw angle of approximately $90°$ had occurred between $t \approx 47769$ s and $t \approx 47775$ s, as indicated by the raw output of the VCS-z gyroscope (*Gyro_RotRate_z*). The rotation rate during the turn is about $20°$/s (the plot's unit is $2^{-17} \cdot 360°/s$ here). During this turn, the field strength in VCS-x, y changes accordingly. A subsequent turn is initiated at $t \approx 47825$ s, after the failure of the magnetometer, which is not reflected in the magnetic field reading anymore. After approximately $120°$ of turning, the wrong compass reading interferes with the attitude filter's on-line calibration badly enough to completely spoil the attitude estimation in all coordinates, which hazardously destabilizes the flight controller's position control layer.

It remains to be clarified why the safety pilot failed to regain control. The last graph, *ServosIn_uav*, depicts the state of the remote control channel selecting between UAV or manual control modes (this is usually 0 or 1, but scaled by 3000 here for visibility). At $t \approx 47835$ s, immediately after the beginning failure of position control, the pilot disengaged UAV mode, but reengaged it a fraction of a second later, thus keeping the impeded flight controller at work. The explanation finally arrived upon is that he accidentally threw the UAV switch on the remote control back while moving his fingers to the transmitter sticks.

This example proves that data logging with low sampling rates is still highly useful, and that FDR-style logging in general is effectively indispensable.

Of course, a failure scenario like this one can – and should – be avoided via *fault detection and identification* (FDI) procedures, see e.g. [71, 69]. Sensor failures like a stuck output or an output's exceeding a certain nominal range of values can easily be detected automatically. There are two possible approaches:

- The consequences of certain failures can be automatically minimized by providing specialized *fails-safe procedures* for system operation in certain foreseen cases of failure. This is covered in more detail in the next section. In order to gain maximum benefit from this kind of measure, it should be implemented in the on-board software, so that the fail-safe mode can be engaged with minimum delay and even in case of communication outage.

- If an FDR-style data log is available at the ground station anyway, FDI measures can also be implemented there. The advantage is that usually more computational power can be spent there for FDI purposes. One promising approach is to run a full simulation of the UAV dynamics and compare its output with the data actually received from the UAV. The disadvantage, however, is that on-ground FDI is primary suitable for informing human operators of some impeding problem. Of course, integrated approaches of on-ground detection and on-board handling of faults are imaginable as well.

Meanwhile, a ground-station based FDI module has been added to the MARVIN system that would have alerted the human operator in a situation like the one presented above. However, it is important to set the alarm thresholds not too low. Otherwise, the FDI module might issue frequent alarms even if no actual failure exists. Then, the human operator is tempted to ignore these warnings, including potentially relevant ones.

One minor recommendation of possible significance is *never* to use proper hard disk drives
on board for data storage. During the COMETS project [28], one of the partners used a 2.5"
laptop hard disk for on-board data storage, but the drive frequently interrupted its operation
whenever it detected external acceleration. Moving the heads into "park" position is a reasonable measure to protect a laptop drive in case of accidental strokes or impacts, but is extremely
annoying when moderate vibration and motion constitute a nominal property of the drive's operating environment.

7.3   Failsafe Procedures

Failsafe procedures shall refer to all operating modes of a UAV system that are only engaged
when certain system components fail to work as desired. In view of the architectural compo-
sition as assumed in this book, such procedures may be implemented in the controller, in the
sensor interfacing code, and/or in the sensor fusion and filtering algorithms.

Sensor-based and filter-based failsafe procedures are highly specialized with respect to the
sensors and filter algorithm in use. Possible failsafe procedures for the MARVIN attitude filter have already been discussed specifically in section 6.3.3 and will not be readdressed here.
In addition, model-based attitude estimators may be generally used to compensate for the failure of drift-compensating sensor groups like accelerometers and magnetometers, probably at
the expense of reduced data quality. In the presence of a working high-performance position
sensor, it may even be possible to replace all attitude-sensing hardware through such an esti-
mator and still maintain autonomous flight of a helicopter-class UAV. It is certainly possible to
perform flight control of an aerodynamically stable fixed-wing UAV solely based on position
data, given that only maneuvers with limited dynamics are performed. However, such control
approaches have not been actually investigated by the author, nor within the author's group.

The remainder of this section concentrates on control-based failsafe procedures. This includes
control modes specifically provided for use during reduced-quality sensor data.

7.3.1   Human Safety Pilot

The best ultimate fallback procedure with any kind of UAV is, of course, manual control by
a human safety pilot. There are several reasons that render this kind of failsafe option highly
desirable:

1. Operation by human remote control usually requires the *smallest number of on-board
   components* to work properly. No sensors might be involved at all. Therefore, the human
   safety pilot is the failsafe procedure applicable to the greatest subset of possible failure
   events.

2. Control by the safety pilot usually involves only *off-the-shelf components*, i.e. components that are not by themselves part of the UAV system design process. Hence, these
   components can be rather considered reliable than those subject to development, and at
   a much earlier time within the development process.

3. The human safety pilot can react *most flexibly* to unforeseen kinds of failure, and successfully adapt to a new kind of failure in the shortest possible time.

4. The safety pilot's *personal experience* with similar remote-controlled vehicles provides a reliable basis for operating the UAV under development. This concerns both the very first flight control tests in an early phase of system development and regular verification of the air vehicle's proper function prior to autonomous operation. Without such experience, initial flight testing would either require a complex mechanical support device to ensure safety or involve a substantial risk of loss of the vehicle.

Especially the last point deserves careful consideration. When a UAV is developed that does not correspond, maybe approximately, to any air vehicle that has already been operated under remote control by human pilots, it is extremely difficult to devise a deterministic procedure of operation for the first flight tests. The use of a mechanical safety support (see e.g. [21]) will usually be mandatory, especially since all flight tests with a vehicle not operatable by humans need to begin with an autonomous takeoff, the most critical phase of autonomous flight there is. A mechanical safety support, unfortunately, does affect the flight dynamics of a UAV in a more or less significant way.

On the other hand, flight tests of UAVs that are operatable by humans can be initiated while the vehicle is airborne at a safe altitude. In case of undesirable behavior, the safety pilot can switch back to remote control and recover the vehicle into a safe state. This kind of testing regime has been successfully applied without exception during all development phases of the MARVIN system, without a single loss of a vehicle subsequent to airborne handover.

Doing so does require flight control to be able to adopt the current operating point as used by the human pilot at the moment of handover. This is a special feature to be implemented in the controller, which can be accomplished by adjusting the integrators (i.e. bias compensation stages according to the approach set forth in chapter 5) without significant difficulty.

However, the activity of recovering a UAV from undesirable behavior of the flight controller is much more challenging than full-time manual flight. This is particularly true for helicopter-class vehicles that need to be actively stabilized by the pilot. For during manual flight, the pilot may always know what control command has caused the currently observed behavior, and critical flight situations do not usually occur. On the other hand, threatening actions performed by the controller must be deduced from observation only, and recovering will frequently occur as a result of a critical flight situation. Therefore, only a very experienced pilot can reliably act as a safety pilot.

7.3.2 Special Control Modes

Apart from full manual remote control, it will often be desirable to provide specialized control modes to be applied in certain situations. These may serve for planned testing purposes, or for coping with sensor failures, as already motivated in section 7.2 above. At least the following classes of special control modes may be considered in general:

Manual high-level control. As indicated through the three control architecture examples in chapter 5, flight control usually consists of two layers, one of which is position control and the other one attitude control. This is due to the fact that with all common air vehicles, lift and propulsion forces need to be rotated together with the vehicle to yield the desired position trajectory. Therefore, the position control layer or parts thereof may be disabled and replaced by manual control inputs. A human pilot may command the desired orientation, or its rate of change, or the desired velocity, for example, via remote control. Then, lower-level or attitude control may work as usual, rendering this kind of manual control much easier than full manual remote-controlled piloting. For unmanned helicopters, anyone may be able to perform this way of operation with very little training, while full manual remote control is a very difficult task for humans. Hence, manual high-level control may be highly beneficial, both for testing purposes and for recovery in cases of high-level sensor failure (e.g. GPS outage).

No high-level control. In case of failure of the position sensor, e.g. temporary GPS outage, upper-level position control cannot be maintained at all, because position estimation with sufficient accuracy is not possible with sensors suitable for small UAV application. In this case, the best-possible solution is to disable position control and to fix the desired attitude of the lower control layer according to the current operating point. This will prevent even helicopter-class UAVs from crashing immediately, but the vehicle may start drifting away faster and faster. Additionally, as explained in detail in section 6.1, the lack of position data will also disable the drift compensation in any attitude measurement based on rotation rate and acceleration sensors, so even attitude control would ultimately fail. Hence, this special control mode is suitable for bridging short-term failures of the position sensor on the one hand, and provides the human safety pilot with a very comfortable reaction time on the other.

Low-gain control. When the accuracy of some sensor or sensor group deteriorates, maybe temporarily, it may be advisable to operate parts of flight control with reduced gain factors, for sensor noise and other kinds of error will have less effect on the system behavior then. While control gains are often critical to stability with traditional controller design procedures, the design approach set forth in chapter 5 is guaranteed to maintain stability for any reduction of the k_j-gains of figure 5.5 and (5.13). Therefore, such a reduced-gain control mode can always be included without having to consider stability. Of course, the geometry of possible flight trajectories will be affected by gain reduction, so it may be advisable to interrupt the execution of the mission until all sensors revert to nominal operation. In the moment of restored accuracy, gains should remain reduced until factual convergence of the system state.

In the MARVIN system, two special modes involving no position control and low-gain position control are provided. The former one is employed in case of total GPS outage for more than one second, the latter one applies when the GPS receiver reports degraded accuracy. This may occur if the receiver fails to evaluate the carrier phase differences, or if the stream of differential correction data from the ground-based reference receiver is interrupted.

7.4 Subsumption

This chapter has mentioned several tools that may facilitate the design and operation of a small UAV system considerably. Among them, software-in-the-loop testing, FDR-style data logging, and a human safety pilot constitute generally recommended system features.

When choosing a type of aerial vehicle to be turned into a UAV, using a human-operatable vehicle must generally be considered preferable to not doing so. As a human safety pilot, UAV designers should spend every effort to employ the best remote control pilot available for the type of vehicle in question.

In addition to the more general safety tools, the provision of special control modes for the most probable kinds of failure will help to minimize the risk of damage, even if they cannot fully eliminate the potential need for human intervention.

Chapter 8

Conclusion

This book has dealt with the requirements and approaches relevant to the design of small UAV systems, in so far as these requirements can be generalized instead of being restricted to a particular mission scenario. Aspects covered in this respect have been on-board computing equipment, communications hardware and software, flight control implementation, on-board sensors, and procedures and tools that ensure safety and help provide diagnoses during the development and application of small UAVs.

Summing up the findings of this book, the most important result is that today, given the tools and methods discussed here, the design of a small UAV system is mainly routine. This is absolutely true for bigger vehicles, say beyond 5 kg take-off mass, for in this case, the design task mainly consists of combining components on a "user level". For smaller vehicles, the aspect of developing special miniaturized components gains more and more emphasis within the design process, which entails a completely different class of qualification: developing intelligent sensors, communication devices, or actuators smaller than commercially available is clearly more challenging – and far more specialized – than the combining of off-the-shelf components.

The issue of size and its implication to possible payload systems is also critical with respect to marketability – today, the smallest-designable UAVs, fixed-wing planes with about 15 cm wingspan, seem not yet attractive commercially because of the limited utility of their possible payloads and their restricted autonomy.

On the other hand, the design of bigger small UAVs, which can be accomplished by combining off-the-shelf components, is clearly more an "art of breadth" than an "art of depth", because the specification sheets supplied with each of the individual components basically contain everything the designer really needs to know. The main challenge, instead, is to overlook the complicated network of requirements and possible solution alternatives well enough to come up with an "optimal" design, for this "optimization" cannot be performed component by component, but only from a global view.

As far as individual findings are concerned, a very rude summary to be supplied at this point of this book consists of methods and components that *may* be surprisingly dispensable – or less important than expected –, and of methods and components that *may* be surprisingly useful – or more important than expected. The former are e.g. acceleration sensors, off-the-shelf operating systems, off-the-shelf middleware, Kalman filters, Lyapunov functions, video cameras, and on-board PCs. The latter include real-time-aware small-footprint middleware, RC receivers by ACT, low-bandwidth data logs, position sensors, and compasses.

Index

Bibliography

[1] *Aero-Tec Helicopter-Technik Uwe Welter.* Oberhausen-Rheinhausen. *http://www.aero-tec-helicopter.de.*

[2] *Canon Digital Imaging Developer Programme. http://www.didp.canon-europa.com.*

[3] *gPhoto^2 Digital Camera Software. http://www.gphoto.org.*

[4] Middleware Resource Center. *http://www.middleware.org.*

[5] RTLinux-gpl FTP repository. *http://www.rtlinux-gpl.org/,* February 2005.

[6] AWARE Project, Platform for Autonomous self-deploying and operation of Wireless sensor-actuator networks cooperating with AeRial objEcts. IST-2006-33579, 2006–2009. *http://grvc.us.es/aware/main.shtml.*

[7] ISO TC 22/SC 3. *ISO 11898-1 Road vehicles – Controller area network (CAN) – Part 1: Data link layer and physical signalling.* International Organization for Standardization, 2003.

[8] ISO TC 22/SC 3. *ISO 11898-2 Road vehicles – Controller area network (CAN) – Part 2: High-speed medium access unit.* International Organization for Standardization, 2003.

[9] ISO TC 22/SC 3. *ISO 11898-4 Road vehicles – Controller area network (CAN) – Part 4: Time-triggered communication.* International Organization for Standardization, 2003.

[10] Höft & Wessel AG. *HW86010 DECT Transceiver Module: The optimum embedded wireless solution.* Hannover. *http://www.hoeft-wessel.com.*

[11] Infineon Technologies AG. *Instruction Set Manual for the C166 Family of Infineon 16-Bit Single-Chip Microcontrollers V2.0.* München, 2001-03 edition, March 2001.

[12] Infineon Technologies AG. *Product Catalog Microcontrollers – Version 11.0.* München, 2006-01 edition, 2006.

[13] Michele Amoretti, Stefano Bottazzi, Stefano Caselli, and Monica Reggiani. Telerobotic Systems Design based on Real-Time CORBA. *Int. Journal of Robotic Systems, Special issue on Internet & Online Robots for Telemanipulation.,* 22(4):183–201, April 2004.

[14] Inc. Analog Devices. *ADXRS300 Data Sheet Rev.B:* ±*300°/s Single Chip Yaw Rate Gyro with Signal Conditioning.* Norwood, MA, 2004. *http://www.analog.com.*

[15] Inc. Analog Devices. *ADXL322 Data Sheet: Small and Thin* ±*2 g Accelerometer.* Norwood, MA, 2005. *http://www.analog.com.*

[16] Communications Assistant Secretary of Defense for Command, Control and Intelligence. *Global Positioning System Standard Positioning Service Performance Standard,* October 2001. *http://www.navcen.uscg.gov/gps.*

[17] 1394 Trade Association. *IIDC 1394-based Digital Camera Specification, Version 1.31, TA Document 2003017,* February 2004. *http://www.1394ta.org.*

[18] Electronics Industries Association. Interface between data terminal equipment and data communication equipment employing serial data interchange. EIA recommended standard RS-232-C, EIA, August 1969.

[19] Michael Barabanov and Victor Yodaiken. Introducing Real-Time Linux. *Linux Journal,* (34), February 1997.

[20] S. S. Beauchemin and J. L. Barron. The computation of optical flow. *ACM Computing Surveys,* 27(3):433–467, 1995.

[21] Markus Bernard, Konstantin Kondak, and Günter Hommel. Framework for development and test of embedded flight control software for autonomous small size helicopters. In Günter Hommel and Sheng Huanye, editors, *Embedded Systems – Modeling, Technology, and Applications, Proceedings of the 7th International Workshop held at Technische Universität Berlin, June 26/27, 2006,* Heidelberg and others, 2006. Springer. ISBN 1-4020-4932-3.

[22] U. W. Brandenburg, M. Finke, D. Hanisch, M. Musial, and R. Stenzel. TUBROB – An autonomously flying robot. In *Proc. Symposium of the Association for Unmanned Vehicle Systems 1995,* pages 32–42, Washington D.C., USA, 1995.

[23] U. W. Brandenburg, M. Finke, and M. Musial. Aufbau und Steuerung des fliegenden Roboters TUBROB. In *Autonome Mobile Systeme, 11. Fachgespräch, Karlsruhe,* pages 100–109, Berlin, Heidelberg, New York, 1995. Springer-Verlag.

[24] M. Buschmann, J. Bange, and P. Vörsmann. MMAV - a miniature unmanned aerial vehicle (mini-UAV) for meteorological purposes. In *16th Symposium on Boundary Layers and Turbulence,* Portland, ME, August 2004. American Meteorological Society. P6.7.

[25] Raja Chatila et al. EDEN: Robotics in Natural Environments. *http://www.laas.fr/~simon/eden/index.php.*

[26] Inc. Cloud Cap Technology. *Crista Sensor Head Interface Control Document.* Hood River, OR, May 2004. *http://www.cloudcaptech.com.*

[27] Speake & Co. *FGM-Series Magnetic Field Sensors.* Llanfapley. *http://www.speakesensors.com.*

[28] COMETS consortium. Real-time coordination and control of multiple heterogeneous unmanned aerial vehicles. IST-2001-34304, 2002–2005. http://www.comets-uavs.org/.

[29] Atmel Corporation. Microcontrollers and DSP – AVR 8-bit RISC. *http://www.atmel.com.*

[30] Atmel Corporation. *ATmega48/88/168 Preliminary Complete Datasheet Revision G.* San Jose, CA, June 2006. *http://www.atmel.com.*

[31] Compaq Computer Corporation, Hewlett Packard Company, et al. *Universal Serial Bus Specification, Revision 2.0,* April 2000. *http://www.usb.org.*

[32] Microsoft Corporation. Mobile Developer Center: Windows CE 5.0 Product Documentation. *http://msdn.microsoft.com.*

[33] NEC TOKIN Corporation. *NEC TOKIN Sensors Vol.03.* Tokyo, March 2006. *http://www.nec-tokin.com.*

[34] PNI Corporation. *TCM2 Electronic Sensor Module User's Guide.* Santa Rosa, CA, February 2004. *http://www.pnicorp.com.*

[35] RadiSys Corporation. Microware OS-9. *http://www.radisys.com.*

[36] Shawn Corwyn Coyle. *The Art and Science of Flying Helicopters.* Iowa State Univ Pr (Trd), February 1996.

[37] Inc. Crossbow Technology. *VG400 Solid-State Vertical Gyro.* San Jose, CA. *http://www.xbow.com.*

[38] Carsten Deeg. *Modeling, Simulation, and Implementation of an Autonomously Flying Robot.* PhD thesis, Technische Universität Berlin, 2006.

[39] Abteilung Systemautomation Deutsches Zentrum für Luft-und Raumfahrt e. V. (DLR). *Autonomes VTOL UAV.* Braunschweig. *http://www.dlr.de.*

[40] Christoph Eck. *Navigation Algorithms with Applications to Unmanned Helicopters.* PhD thesis, ETH Zürich, 2001. Diss. ETH Nr. 14402.

[41] Deborah Estrin, Ramesh Govindan, John Heidemann, and Satish Kumar. Next century challenges: scalable coordination in sensor networks. In *MobiCom '99: Proceedings of the 5th annual ACM/IEEE international conference on Mobile computing and networking,* pages 263–270, New York, August 1999. ACM Press. ISBN 1-58113-142-9.

[42] European Telecommunications Standards Institute (ETSI). *Digital Enhanced Cordless Telecommunications (DECT); Common Interface (CI); Part 1: Overview; ETSI Doc. Number EN 300 175-1,* June 1999. *http://www.etsi.org.*

[43] European Telecommunications Standards Institute (ETSI). *Digital cellular telecom-munications system (Phase 2+) (GSM); General Packet Radio Service (GPRS); Service description; Stage 2 (3GPP TS 03.60 version 6.8.0 Release 1997)*, March 2001. *http://www.etsi.org.*

[44] International Organization for Standardization. *International Standard ISO/IEC 7498-1: Information Technology – Open Systems Interconnection – Basic Reference Model: The Basic Model*, second edition, June 1996.

[45] Association for Unmanned Vehicle Systems International. *http://www.auvsi.org/.*

[46] Association for Unmanned Vehicles International. *http://avdil.gtri.gatech.edu/AUVS/IARCLaunchPoint.html.*

[47] DECT Forum. *DECT – The standard explained*, February 1997. *http://www.dect.org.*

[48] Free Software Foundation. GCC (GNU Compiler Collection) Homepage. *http://gcc.gnu.org/.*

[49] Free Software Foundation. GNU Binutils. *http://www.gnu.org/software/binutils/.*

[50] Inc. Freescale Semiconductor. *MC68332 Technical Summary*, 1996. Document order number MC68332TS/D Rev. 2.

[51] Inc. Freescale Semiconductor. *MC68F375 Reference Manual Revision 25*, June 2003.

[52] Inc. FreeWave Technologies. *Ethernet Series: FGR-HTPlus Industrial Radio, Specifications.* Boulder, CO. *http://www.freewave.com.*

[53] Inc. FSMLabs. RTLinuxPro.

[54] John Geen and David Krakauer. New iMEMS angular-rate-sensing gyroscope. *Analog Dialogue, ADI Micromachined Products Division*, 37(03), 2003. *http://www.analog.com.*

[55] Hans Peter Geering et al. Unmanned Aerial Vehicle Group. *http://www.uav.ethz.ch.*

[56] Michael George and Salah Sukkarieh. Tightly coupled ins/gps with bias estimation for uav applications. In *Proceedings Australasian Conference on Robotics and Automation 2005.* ARAA (Australian Robotics and Automation Association), 2005. *http://www.araa.asn.au.*

[57] CAPTRON Electronic GmbH. *HeliCommand-Profi.* München. *http://www.helicommand.com.*

[58] Robert Bosch GmbH. *CAN Specification Version 2.0.* Stuttgart, 1991.

[59] Greg Goebel. Unmanned aerial vehicles, v1.4.0. *http://www.vectorsite.net/twuav.html*, April 2006.

[60] D. Gordon and R. Brown. Recent advances in fluxgate magnetometry. *IEEE Transactions on Magnetics*, 8(1):76–82, March 1972. ISSN 0018-9464.

[61] J. M. Grasmeyer and M. T. Keennon. Development of the black widow micro air vehicle. In *AIAA Paper No. AIAA-2001-0127, Proc. 39th AIAA Aerospace Sciences Meeting and Exhibit*. American Institute of Aeronautics and Astronautics, Inc., 2001.

[62] Object Management Group. Minimum corba specification, version 1.0. *http://www.omg.org/*, August 2002.

[63] Object Management Group. Realtime-CORBA specification, version 2.0. *http://www.omg.org/*, November 2003.

[64] Object Management Group. Common object request broker architecture: Core specification, version 3.0.3. *http://www.omg.org/*, March 2004.

[65] Object Management Group. Event service specification, version 1.2. *http://www.omg.org/*, October 2004.

[66] Object Management Group. Notification service specification, version 1.1. *http://www.omg.org/*, October 2004.

[67] Object Management Group. Real-time CORBA specification, version 1.2. *http://www.omg.org/*, January 2005.

[68] Real-Time Systems Group. MARVIN – An autonomously operating flying robot. *http://pdv.cs.tu-berlin.de/MARVIN*. TU Berlin.

[69] Rajesh Mahtani Guillermo Heredia, Aníbal Ollero, Manuel Béjar, Volker Remuß, and Marek Musial. Detection of sensor faults in autonomous helicopters. In *Proc. of the 2005 IEEE International Conference on Robotics and Automation (ICRA 2005), Barcelona, Spain, April 2005*, Barcelona, Spain, April 2005.

[70] J. Hanisch, P. Ergenzinger, and M. Bonte. Dumpling - an "intelligent" boulder for studying internal processes of debris flows. In Dieter Rickenmann and Cheng-Lung Chen, editors, *Proceedings of the Third International Conference on Debris-Flow Hazards Mitigation*, September 2003. ISBN 90-77017-78-X.

[71] G. Heredia, V. Remuß, A. Ollero, R. Mahtani, and M. Musial. Actuator fault detection in autonomous helicopters. In *Proc. of the 5th IFAC Symposium on Intelligent Autonomous Vehicles (IAV 2004)*, Lisboa, July 2004. Elsevier Science. ISBN 008-044237-4.

[72] Siemens Home and Office Communication Devices GmbH & Co. KG. *Gigaset M101 Data – The cordless V.24/RS232 interface for PCs, modem and other equipment – Operating instructions*, 2006. Download version: 2.057, *http://gigaset.siemens.com*.

[73] DraganFly Innovations Inc. *DraganFlyer 5 Ti Manual*. Saskatoon, Canada. *www.rctoys.com*.

[74] MicroStrain Inc. *3DM-GX1 Gyro Enhanced Orientation Sensor*. Williston, VT, 2005. *http://www.microstrain.com*.

[75] NovAtel Inc.   *OEMV-2 Data Sheet, Version 1B*.   Calgary, Alberta, 2006. *http://www.novatel.com*.

[76] *IEEE Standard for Information Technology – LAN/MAN – Specific requirements – Part 11: Wireless LAN Medium Access Control (MAC) and Physical Layer (PHY) specifications*, 1999. IEEE Standard No.: 802.11 / 8802-11.

[77] Institute of Electrical and Electronics Engineers. *Portable Operating System Interface (POSIX), IEEE Standard No.: 1003.1-2001*, 2001. ISBN: 0-7381-3010-9, IEEE Product No.: WE94956.

[78] Eric N. Johnson et al.   Georgia   Tech   UAV   Research   Facility. *http://controls.ae.gatech.edu/uavrf*.

[79] Eric N. Johnson and Suresh K. Kannan. Adaptive trajectory control for autonomous helicopters. *jgcd*, 28(3), May-June 2005.

[80] Wayne Johnson. *Helicopter Theory*. Dover Publications, Inc. New York, 1994.

[81] John D. Anderson Jr. *Introduction to Flight*. McGraw-Hill Book Co., second edition, 1985.

[82] R. E. Kalman. A new approach to linear filtering and prediction problems. *Transactions of the ASME Journal of Basic Engineering*, 82:35–45, 1960.

[83] R.E. Kalman and R.S. Bucy. New results in linear filter and prediction theory. *Journal of Basic Engineering*, pages 95–108, 1961.

[84] Graupner GmbH & Co. KG. *Gyro System SRVS G490T (Order No. 5137) Operating Manual*. Kirchheim/Teck, May 2003.

[85] Hassan K. Khalil. *Nonlinear Systems*. Prentice Hall, 3rd edition, 2001. ISBN 0-130-67389-7.

[86] H. J. Kim, D. H. Shim, and S. Sastry. Flying robots: Modeling, control, and decision making. In *International Conference on Robotics and Automation 2002*, May 2002. *http://robotics.eecs.berkeley.edu/bear/publications.html*.

[87] T. Kordes, M. Buschmann, S. Winkler, H.-W. Schulz, and P. Vörsmann. CAROLO – Entwicklung eines autonomen Mikroflugzeugs an der TU Braunschweig. In *Deutscher Luft- und Raumfahrtkongress 2003, München*, Bonn, 2003. Deutsche Gesellschaft für Luft- und Raumfahrt - Lilienthal-Oberth e.V. (DGLR).

[88] Miroslaw Krstić, Ioannis Kanellakopoulos, and Petar Kokotović. *Nonlinear and Adaptive Control Design*. John Wiley & Sons, Inc., New York and others, 1995. ISBN 0-471-12732-9.

[89] EDEN Project LAAS/CNRS. *Karma, The Blimp*. Toulouse. *http://www.laas.fr*.

[90] J. Gordon Leishman. *Principles of Helicopter Aerodynamics*. Cambridge University Press, 2000.

[91] Qing Li and Caroline Yao. *Real-Time Concepts for Embedded Systems*. CMP Books, 1st edition, July 2003. ISBN 1-57820-124-1.

[92] ARM Ltd. ARM – the architecture for the digital world. *http://www.arm.com*.

[93] Datatyker Pty Ltd. *dataTaker 905U-D High Speed Industrial Radio Modem Specifications*. Melbourne. *www.datataker.com*.

[94] Inc. Lynx Real-Time Systems. LynxOS Real-Time Operating System. *http://www.lynx.com/*.

[95] Maxim Integrated Products. *MAX3110E/MAX3111E SPI/MICROWIRE-Compatible UART and ±15kV ESDProtected RS-232 Transceivers with Internal Capacitors*, July 1999.

[96] Sergio Montenegro, Stefan Jähnichen, and Olaf Maibaum. Simulation-based testing of embedded software in space applications. In Günter Hommel and Sheng Huanye, editors, *Embedded Systems – Modeling, Technology, and Applications, Proceedings of the 7th International Workshop held at Technische Universität Berlin, June 26/27, 2006*, Heidelberg and others, 2006. Springer. ISBN 1-4020-4932-3.

[97] 1 Inc. Motorola. *M68HC11 Microcontrollers Revision 4.1*, February 2002. *http://www.motorola.com/semiconductors*.

[98] Inc. Multi-Tech Systems. *MultiModem GPRS External Wireless Modem Datasheet*. United Kingdom, 2005. *http://www.multitech.com*.

[99] M. Musial, U. W. Brandenburg, and G. Hommel. MARVIN – Technische Universität Berlin's flying robot for the IARC Millennial Event. In *Proc. Symposium of the Association for Unmanned Vehicle Systems 2000*, Orlando, Florida, USA, 2000.

[100] Marek Musial, Uwe Wolfgang Brandenburg, and Günter Hommel. Cooperative Autonomous Mission Planning and Execution for the Flying Robot MARVIN. In Enrico Pagallo, Frans Groen, Tamio Arai, et al., editors, *Intelligent Autonomous Systems 6*, pages 636–646, Amsterdam, Berlin, Oxford, et al., 2000. IOS Press and Ohmsha. ISBN 1-58603-078-7.

[101] Marek Musial, Uwe Wolfgang Brandenburg, and Günter Hommel. MARVINs Sieg im "Millennial Event" - Erfolg durch minimale Lösungen. In R. Dillmann, H. Wörn, and M. v. Ehr, editors, *Autonome Mobile Systeme 2000, 16. Fachgespräch, Karlsruhe*, pages 328–336, Berlin, Heidelberg, New York, 2000. Springer-Verlag. ISBN 3-540-41214-X.

[102] Marek Musial, Uwe Wolfgang Brandenburg, and Günter Hommel. Inexpensive system design: The flying robot MARVIN. In *Unmanned Air Vehicle Systems: Sixteenth International UAVs Conference*, pages 23.1–23.12, Bristol, UK, 2001. University of Bristol. ISBN 0-86292-517-7.

[103] Marek Musial, Uwe Wolfgang Brandenburg, and Günter Hommel. MARVIN siegt im "'Millennial Event'". *Physik in unserer Zeit*, 33(1):30–36, 2002. ISSN 0031-9252.

[104] Marek Musial, Carsten Deeg, Volker Remuß, and Günter Hommel. Communication system for cooperative mobile robots using ad-hoc networks. In *Proc. of the 5th IFAC Symposium on Intelligent Autonomous Vehicles (IAV 2004)*, Lisboa, July 2004. Elsevier Science. ISBN 008-044237-4.

[105] Marek Musial, Carsten Deeg, Volker Remuß, and Günter Hommel. Orientation sensing for helicopter uavs under strict resource constraints. In *Proc. First European Micro Air Vehicle Conference EMAV 2004*, Braunschweig, Germany, July 2004. German Institute of Navigation (DGON).

[106] Koninklijke Philips Electronics N.V. *LPC2104/2105/2106 Single-chip 32-bit microcontrollers*, October 2003. Document order number 9397 750 12142.

[107] Institute of Electrical and Electronics Engineers. *Test Access Port and Boundary-Scan Architecture*. IEEE, Piscataway, N.J., January 1992. IEEE Standard 1149.1-1990.

[108] Institute of Electrical and Electronics Engineers. *IEEE Standard for Information technology – Telecommunications and information exchange between systems – Local and metropolitan area networks – Specific requirements – Part 3: Carrier Sense Multiple Access with Collision Detection (CSMA/CD) Access Method and Physical Layer Specifications*, 2002. IEEE Standard No.: 802.3.

[109] Institute of Electrical and Electronics Engineers. *IEEE Standard for Information technology – Telecommunications and information exchange between systems – Local and metropolitan area networks – Specific requirements – Part 11: Wireless LAN Medium Access Control (MAC) and Physical Layer (PHY) specifications – Amendment 4: Further Higher-Speed Physical Layer Extension in the 2.4 GHz Band*, 2003. IEEE Standard No.: 802.11g.

[110] Scott Pace, Gerald P. Frost, Irving Lachow, et al. *The Global Positioning System: Assessing National Policies*. RAND Corporation, Santa Monica, CA, 1995. ISBN 0-8330-2349-7.

[111] Honeywell Sensor Products. *1- and 2-Axis Magnetic Sensors HMC1001/1002, HMC1021/1022 Data Sheet Rev. B.* Plymouth, MN, April 2000. *http://www.ssec.honeywell.com.*

[112] Honghui Qi and J. B. Moore. Direct Kalman filtering approach for GPS/INS integration. *IEEE Transactions on Aerospace and Electronic Systems*, 38(2):687–693, April 2002. ISSN 0018-9251.

[113] Volker Remuß, Marek Musial, and Uwe Wolfgang Brandeburg. BBCS – Robust communication system for distributed systems. In Thomas Christaller et al., editors, *Proc. IEEE Internat. Workshop on Safety, Security and Rescue Robotics (SSRR 2004)*, May 2004. ISBN 3-8167-6556-4.

[114] Volker Remuß, Marek Musial, Carsten Deeg, and Günter Hommel. Embedded system architecture of the second generation autonomous unmanned aerial vehicle MARVIN MARK II. In Günter Hommel and Sheng Huanye, editors, *Embedded Systems – Modeling, Technology, and Applications, Proceedings of the 7th International Workshop held at Technische Universität Berlin, June 26/27, 2006*, Heidelberg and others, 2006. Springer. ISBN 1-4020-4932-3.

[115] Volker Remuß, Marek Musial, and Günter Hommel. Marvin – an autonomous flying robot based on mass market components. In *Proc. Workshop WS6 Aerial Robotics, International Conference on Intelligent Robots and Systems (IROS2002)*, pages 23–28, Lausanne, Switzerland, 2002. IEEE/RS.

[116] Arthur Alexander Reyes et al. AVL: Autonomous Vehicles Laboratory. *http://www3.uta.edu/faculty/reyes/AVL/Default.htm*.

[117] Wind River. VxWorks 6.2 product note. *http://www.windriver.com*, 2006.

[118] S. Shankar Sastry et al. BEAR: Berkeley Aerobot Team. *http://robotics.eecs.berkeley.edu/bear/*.

[119] Wolfgang Schäper and Stephan Lämmlein. Remote controlled vehicles for meteorological measurements in the lower atmosphere. In *Proc. First European Micro Air Vehicle Conference EMAV 2004*, Braunschweig, Germany, July 2004. German Institute of Navigation (DGON).

[120] Conrad Electronic SE. *Drahtlose Farb-Micro-Pinhole-Kamera Artikel-Nr. 751176–62*. Hirschau, 2006. *http://www.conrad.de*.

[121] Conrad Electronic SE. *GPS Receiver 16 Kanal CR4 (USB) Artikel-Nr. 989777–62*. Hirschau, 2007. *http://www.conrad.de*.

[122] David Seal, editor. *ARM Architecture Reference Manual Second Edition*. Addison-Wesley, 2000. ISBN 0-201-73719-1, ARM Doc No.: DDI-0100.

[123] Philips Semiconductors. *The I^2C-Bus Specification Version 2.1*, January 2000. *http://www.semiconductors.philips.com*.

[124] Inc. SensComp. *600 Series Smart Sensor Data Sheet*. Livonia, MI, September 2004. *http://www.senscomp.com*.

[125] D. H. Shim, H. J. Kim, and S. Sastry. Control system design for rotorcraft-based unmanned aerial vehicles using time-domain system identification. In *Proceedings of the 2000 IEEE International Conference on Control Applications*, pages 808–813, 2000. ISBN: 0-7803-6562-3.

[126] D. H. Shim, H. J. Kim, and S. Sastry. A flight control system for aerial robots: Algorithms and experiments. In *IFAC Control Engineering Practice*, 2003.

[127] IEEE Computer Society. *IEEE sTandard for a High Performance Serial Bus, IEEE Standard 1394-1995*, August 1996. E-ISBN: 0-7381-1203-8.

[128] QNX Software Systems. QNX Neutrino RTOS. *http://www.qnx.com*.

[129] Andrew S. Tanenbaum. *Modern Operating Systems*. Prentice Hall, 2nd edition, 2001.

[130] O. Tanner and H. P. Geering. Two-degree-of-freedom robust controller for an autonomous helicopter. In *Proceedings of the 2003 American Control Conference*, volume 2, pages 993–998. IEEE, 2003. ISBN 0-7803-7896-2.

[131] Oliver Tanner. *Modelling, Identification, and Control of Autonomous Helicopters*. PhD thesis, ETH Zürich, 2003. Diss. ETH Nr. 14899.

[132] Inc. The MathWorks. MATLAB & Simulink. *http://www.mathworks.com/*.

[133] Dinsmore Sensor Div. The Robson Company, Inc. *DINSmore Sensing Systems General Information*. Girard, PA, October 2003. *http://www.robsonco.com*.

[134] F. Thielecke, J. Dittrich, and A. Bernatz. ARTIS - Ein VTOL UAV Demonstrator. In *Deutscher Luft- und Raumfahrtkongress 2004, Dresden*, Bonn, 2004. Deutsche Gesellschaft für Luft- und Raumfahrt - Lilienthal-Oberth e.V. (DGLR). ISSN 0700-4083.

[135] u-blox AG. *LEA-4A, LEA-4H, LEA-4M, LEA-4P, LEA-4R, LEA-4S, LEA-4T ANTARIS 4 GPS Modules Data Sheet*. Zürich, December 2006. *http://www.u-blox.com*.

[136] International Telecommunication Union. List of definitions for interchange circuits between data terminal equipment (DTE) and data circuit-terminating equipment (DCE). ITU-T Recommendation V.24, ITU-T Telecommunication Standardization Sector, 2000.

[137] Peter Vörsmann et al. Micro Aerial Vehicles. *http://www.ilr.ing.tu-bs.de/forschung/mav/index-en.html*.

[138] weControl. wePilot1000 - A Flight Control System for Remote-Controlled Helicopters. *http://www.wecontrol.ch*.

[139] Martin F. Weilenmann. *Robuste Mehrgrössen-Regelung eines Helikopters*. PhD thesis, ETH Zürich, 1994. Diss. ETH Nr. 10890.

[140] Klaus Westerteicher. *ACT Europe Home*. Pforzheim. *http://www.acteurope.com*.

[141] Wikipedia. Ryan Model 147 Lightning Bug. *http://en.wikipedia.org/wiki/Ryan_Model_147_Lightning_Bug*.

[142] S. Winkler. Das Mikroflugzeug-Projekt CAROLO. In *UAV-/UCAV-/MAV-Aktivitäten in Deutschland, Bremen 21./22. April 2004*, Bonn, 2004. Deutsche Gesellschaft für Luft- und Raumfahrt - Lilienthal-Oberth e.V. (DGLR). *http://www.dglr.de*.

[143] Victor Yodaiken. *FSMLabs Lean POSIX for RTLinux*. FSMLabs, Inc., Socorro, NM, 2000.

www.ingramcontent.com/pod-product-compliance
Lightning Source LLC
LaVergne TN
LVHW062313060326
832902LV00013B/2197